Electrical Installation:
Questions and answers

Second Edition

3.95

This book to be returned on or before
the last date stamped below.

| 9 JUL 1986 | | |

Electrical Installation:
Questions and answers

Michael Neidle
AMIEE, TEng(CEI), FITE, ASEE(Dipl)

Second Edition

Pitman

PITMAN BOOKS LIMITED
128 Long Acre, London WC2E 9AN

Associated Companies
Pitman Publishing New Zealand Ltd, Wellington
Pitman Publishing Pty Ltd, Melbourne

© M Neidle 1982

First published in Great Britain 1975
Second edition 1982
Reprinted 1983

All rights reserved. No part of this publication may be reproduced,
stored in a retrieval system, or transmitted, in any form or by any
means, electronic, mechanical, photocopying, recording and/or
otherwise without the prior written permission of the publishers.
This book may not be lent, resold, hired out or otherwise disposed of
by way of trade in any form of binding or cover other than that in
which it is published, without the prior consent of the publishers.
This book is sold subject to the Standard Conditions of Sale
of Net Books and may not be resold in the UK below the net price.

Printed and bound in Great Britain
at The Pitman Press, Bath

ISBN 0 273 01723 3

Foreword

The second edition of *Electrical Installation: Questions and Answers* underlines the fact that there is a continuous development in electrical installation practice, on the one hand to provide even greater safety for users of a potentially hazardous form of energy and on the other hand as a result of the development of new materials, equipment and techniques which simplify or improve those previously used.

Electrical installation design is often considered as an unimportant branch of electrical engineering, but it should be remembered that the electrical installation forms an essential link between the source of energy and the point of utilisation. Without the installation, most electrical equipment could not be operated at all.

Installations must be correctly designed and constructed to prevent danger of electric shock to persons and livestock and to prevent risk of fire due to electrical faults. The IEE Wiring Regulations, BSI Codes of Practice and various Regulations are necessarily written in legal or quasi-legal language which may not at first sight be clear to many students, operatives and even designers of electrical installations; a book like *Electrical Installation: Questions and Answers* is therefore useful because it presents essential information in a manner which can be readily understood.

C.D. Kinloch DFH
Assistant Director (Technical)
National Inspection Council for
Electrical Installation Contracting (NICEIC)
Member IEE Wiring Regulations Committee

Preface

Electrical installations cover almost every aspect of our national life—domestic, recreational, medical, commercial and industrial. The wiring industry is one which is *progressive and continually expanding*, present annual turnover being some £500 million. Electric wiring has a scientific basis in addition to obvious utilitarian aspects. Installations may in fact be defined as some of the practical branches which are concerned with the utilization and safe application of electricity. It is therefore not surprising that many of the 80 000 people engaged in the industry find it of absorbing interest.

The book is designed for a wide class of readership, covering the requirements of the operative—from apprentice to the old hand—students and users. The present volume should also appeal to workers in allied fields who require a simple, yet authoritative, and thoroughly modern approach to the subject. It will also provide essential background knowledge to the City & Guilds Electrical Installation Courses in addition to relevant sections in the 200 Series.

The question-and-answer method is intended to serve as a break from that adopted in most written works. With this more intimate approach it is hoped to produce a clearer and livelier presentation. However, undue brevity has been avoided, and certain answers have been elaborated where a somewhat deeper treatment is merited.

M. Neidle

Contents

	Foreword	v
	Preface	vi
	Units, Abbreviations, and Symbols	viii
1	Requirements for Safety	1
2	Cables and Flexibles	15
3	Steel-Conduit Wiring Systems and Circuits	20
4	Insulated-Conduit Wiring Systems	34
5	Trunking and Ducting Systems	41
6	Mineral-Insulated Cables	52
7	House and Other Wiring	61
8	The Ring Circuit	73
9	Installation Accessories	80
10	Testing	93
11	Environmental Installations	102
	Index	111

Units, Abbreviations and Symbols

	Unit		Symbol
Voltage	Volt	(V)	V
Current	Ampere	(A)	I
Resistance	Ohm	(Ω)	R
Power	Watt	(W)	P
Energy	Joule	(J)	W
	Kilowatt-hour	(kWh)	W

Multiples and Sub-multiples

Example

Kilo	(k)	= 1000	= 10^3	1 kV = 1000 V or 10^3 V
Mega	(M)	= 1 000 000	= 10^6	1 MΩ = 1 000 000 Ω or 10^6
Milli	(m)	= $\frac{1}{1000}$	= $\frac{1}{10^3}$ = 10^{-3}	1 mA = 10^{-3} A
Micro	(μ)	= $\frac{1}{1\,000\,000}$	= $\frac{1}{10^6}$ = 10^{-6}	1 μΩ = 10^{-6} Ω

Chapter 1
Requirements for Safety

What are the IEE Wiring Regulations?

The full title is the 'Regulations for Electrical Installations' as issued by the Institution of Electrical Engineers and comprise the recognized rules for wiring. Furthermore, a high standard of safety may be assumed for any installation which complies with these Regulations, as they are designed to ensure freedom from fire and electric shock.

Are the IEE Regulations compulsory?

This cannot be answered simply. These Regulations consist of two parts. Part 1 gives the basic requirements for safety and consists largely of extracts from certain statutory regulations, e.g. Electricity Supply Regulations, 1937 and Electricity (Factories) Special Regulations, 1908 and 1944, *and may therefore be taken as compulsory*. Part 2, by far the largest section, details means of compliance with the first part. If Part 1 is not complied with, the electricity supply authority could refuse to connect the installation to the main supply.

What are the main requirements for preventing danger to electrical installations?

(1) Sound workmanship and use of proper materials.
(2) Conductors and apparatus to be of adequate sizes for the currents they are expected to carry, and suitably constructed, installed and protected.
(3) All single-pole switches to be inserted in live, not neutral, conductors.
(4) All conductors to be adequately insulated and mechanically protected.
(5) The metalwork associated with current-carrying conductors to be connected to earth.

(6) Every circuit to be protected by fuse or circuit-breakers so as to prevent danger should excess current flow through it.

What is a fuse?

A fuse is one of the major forms of protection and comes in various forms. A rewirable fuse (Fig. 1 (a)) consists of a thin wire housed in a suitable holder and may press against an asbestos mat or be enclosed in

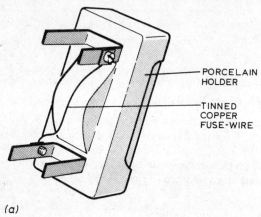

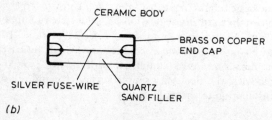

Fig. 1   (a) Typical rewirable fuse
           (b) Cartridge fuse

an asbestos tube. In the cylindrical cartridge type (Fig. 1 (b)) the fuse-wire is hermetically sealed and, as it is not exposed to the atmosphere, the fuse element will not suffer from the effects of corrosion. Additional advantages are:

(1) Fuse element does not deteriorate with age.
(2) Exact calibration.
(3) Rapid operation.

A fuse may be regarded as a 'weak link in the chain'. The fuse-wire is placed in series with the main circuit wiring. In the event of an overload or a 'short-circuit' fault causing an excess of current, the fuse-wire will melt or 'blow', thereby rendering the circuit dead. Short-circuiting takes place when there is a direct connection between the live and neutral conductors, and may occur above a lampholder (Fig. 2). A simple

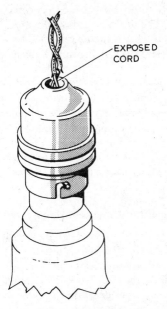

Fig. 2   Short-circuit of flexible cord directly above lampholder

example shows the high current (I) which results because of a 'short'. If, under these circumstances, the resistance (R) to the supply is as low as 0·05 ohm (Ω), the short-circuit current can be calculated by Ohm's law:

$$I = \frac{V}{R}$$

$$= \frac{240 \text{ V}}{0\cdot 05 \Omega}$$

$$= 4\,800 \text{ A}$$

This is an enormous current flow, considering that the standard 3-kilowatt fire takes a current flow of 12·5 amperes (A). Thus it can be understood that such a high current leads to an excessive heat rise which, if adjacent to inflammable materials, will almost certainly cause an outbreak of fire.

A basic safety requirement is that the rating of a fuse shall not exceed the current rating of cables protected by it. Are there any exemptions?

The IEE Wiring Regulations do in fact permit a number of exemptions, four of which are:

(1) 0·5-mm^2 flexible cord (rated at 3 A) as protected by a distribution board 5-A fuse. The flexible cord would almost certainly not supply a lamp rated above 3 A, i.e. above 3 A x 240 V or 720 watts.

(2) Motor circuits. Here the starting surge current exceeds the running current.

(3) 2·5-mm^2 p.v.c. cable as run in conduit may be rated at 21 A. The 30-A ring-circuit fuse provides cable protection due to *diversity*, i.e. all the loads on the ring are unlikely to be switched on together. In any case the ring-circuit arrangement limits the cable loading.

(4) The rating of wiring within enclosures of certain appliances is permitted to be lower than the rating of the load fuse. Any overloading will limit the extent of faulty cables to within the appliance enclosure.

What is a miniature circuit-breaker (m.c.b.)?

The m.c.b., although a somewhat complicated piece of mechanism, is gradually usurping the place of the fuse; it is being fitted in increasing numbers in distribution boards and consumer control units. Its action is based on the principle of the circuit-breaker as fitted to control large industrial circuits, whereby a pre-determined excess of current operates an electromagnet. This in turn, through a switch mechanism, automatically permits the breaking current to be lower than an equivalently rated rewirable fuse.

(1) It can be made to give a close degree of small excess current protection. This is in direct contrast to the rewirable fuse which may be easily wired with a heavier-gauge fuse-wire so that a larger current will flow before the wire actually blows.

(2) If necessary for certain purposes, the mechanism may be fitted with a time-delay action.

(3) After a fault in a circuit has been cleared, the supply can very easily be restored by the simple toggle switch.

How does earthing make for safety?

At the power station or sub-station supply point, the neutral conductor is connected or bonded to earth, so that at this point they are at the same potential. This arrangement forms one of the main electrical safety precautions in Britain. Thus if a live conductor inadvertently touches earthed metal it becomes, in effect, a short-circuit so that the fuse will blow and the circuit is rendered dead. This effect may easily be shown by again applying Ohm's law:

$$I \text{ (current)} = \frac{V \text{ (voltage)}}{R \text{ (resistance)}}.$$

For example, if the resistance of the earth path is 0·5 Ω and a local earth fault occurs, then on a 240-V circuit the current flowing through the live conductors would attempt to rise to 240/0·5 = 480 A. Thus any household fuse-wire would clearly melt, or the miniature circuit-breaker open the circuit, in a very short period of time. In practice the impedance (Z) of the circuit will have to be taken, instead of simply the resistance, but there is no difference in principle.

Reduced voltage is one of the factors making for safety. How should a 110-V electric drill be connected to a 240-V a.c. supply?

Referring to Fig. 3 it will be seen that the transformer supplies a step-down voltage of 110 V to the drill. The secondary winding is centre-tapped to earth so that although the drill operates at 110 V, its maximum

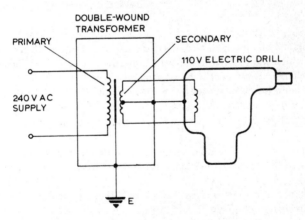

Fig. 3   110-V drill connected to 240-V supply

voltage to earth is only half of this value, namely 55 V. It is essential for safety that the drill casing be properly connected to earth. An alternative method of protection is provided by *double insulation*, where the casing would be constructed of two distinct layers of a durable insulating material.

What is meant by a 'switched neutral'?

For common electric-light control and 13-A switched socket outlets, single-pole switches (Fig. 4) are generally employed. *The switch must be connected to the live side of the supply.*

Fig. 4   Single-pole switch

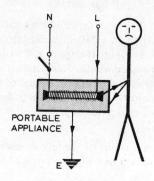

Fig. 5   Incorrect connection by crossed polarity creates a severe potential electric hazard

Crossed polarity or the inadvertent interchange of the L and N conductors may be lethal. The danger of such a situation is illustrated in Fig. 5 which shows what could, and sometimes does, occur when an exposed element is being cleaned or otherwise touched by hand whilst the other hand is in contact with the metal casing. Double-pole switching (Fig. 6) could eliminate this hazard, although some people advocate complete withdrawal of the plug, where possible, as a certain safe form of action.

With screwed-type lampholders, the neutral conductor must be connected to the inner threaded portion and not to the centre contact terminal.

Interchange of the neutral and earth also brings its dangers, as shown by a Government report on electrical fatalities:

A plug on a lead, supplying a caravan from a house, was wired with the *neutral terminal connected to the earth wire and vice versa.* Thus the current taken by the caravan load returned through the earth circuit of the installation which included a section of water-pipe. The deceased, a plumber, took out a section of the pipe, breaking the circuit, and received a fatal shock by touching the remaining section of water pipe. This is a classical case of accidents in countries where water-pipes are used extensively in domestic premises.

What are typical electric-shock values?

Electric currents through the cardiac region may produce dangerous electric shocks especially if sustained for any length of time. The following values are taken from a British Safety Council Report, which further indicates that even these figures may be too high:

Current (mA)

- 1–3   Threshold of perception, generally not dangerous.
- 10–15   Tightening of muscles and difficulty in releasing any object gripped. Acute discomfort.
- 25–30   Extension of muscular tightening to the muscles of the thorax. Dangerous if not quickly stopped.
- Over 50 Fibrillation of the heart which is generally lethal if immediate specialist attention is not given.

Fibrillation of the heart is a major cause of death by electric shock and is caused by uncontrolled and irregular contraction of the heart.

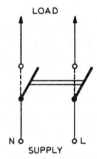

Fig. 6   Double-pole switch

If on entering a room a person is found unconscious and holding a cable, what action should be taken?

Where an isolating switch is at hand it should be switched off immediately to render the circuit dead, otherwise the person must be pulled away from the cable by means of his clothing. Under no circumstances must contact be made with bare flesh as the rescuer could receive a similar shock. Having released the hold from the cable, artificial respiration should be started immediately.

The mouth-to-mouth artificial respiration method should be adopted, this now being considered the best. It requires the patient to be quickly laid on his back with the head tilted back; his nostrils should be pinched and then breathing into his mouth commenced. This action inflates the chest, and when the breath is withdrawn it deflates. The cycle should continue 10–12 times each minute until the arrival of a doctor.

What material is generally used for cable insulation?

P.V.C. (polyvinyl chloride) is commonly used because, when mixed with a plasticizer for flexibility, it is non-hygroscopic, tough, durable, resistant to corrosion and chemically inert and therefore eminently suitable for general wiring work. Further advantages are its resistance to dampness, most acids, alkalis and oils.

What are the limitations to the use of p.v.c.?

Mainly temperature: heat affects the flexibility due to the loss of the plasticizer and consequently p.v.c. cables are limited to installations where the operating temperature does not exceed 70°C. Above the normal working temperature there is a great danger that any pressure or movement would make the conductor strands cut into the insulation and cause decentralization, even to the extent of the possible hazard of short-circuit. Furthermore, chemical changes in the insulation due to high temperatures produce the side-effect of conductor corrosion. This is particularly noticeable on flexible cords where the fine strands may be easily attacked and complete fracture could occur.

At the other end of the scale, general-purpose p.v.c. must not be installed in refrigerated or other situations where the temperatures are consistently below 0°C.

What are the current ratings for metric cables and how do they relate to voltage-drop calculations?

Certain values are given in an IEE Table, of which the following is an abstract, for unenclosed twin-sheathed cables clipped to the surface:

Nominal cross-sectional area (mm^2)	Current rating (amperes)	Voltage drop (mV) per ampere/per metre (mV/A/m)
1·0	12	40
1·5	15	27
2·5	21	16
4·0	27	10
6·0	35	6·8
10·0	48	4·0
16·0	64	2·6

Lighting cables would consist of 1·0 or 1·5 mm^2 cables and at these current ratings overloading is unlikely to occur.

The voltage drop for 25 metres of twin p.v.c.-sheathed cable when carrying 38 A can be calculated by:

$$\text{Voltage drop} = \frac{\text{Actual current carried} \times \text{mV/A/m} \times \text{length of run (metres)}}{1\,000}$$

$$= \frac{38 \times 4 \times 25}{1\,000} = 3 \cdot 8 \text{ V}$$

Incidentally, this value is well within the maximum permitted voltage drop of 2½ per cent of the supply voltage (2½ per cent of 240 V), i.e. 6 V. In practice the type of excess-current protection must also be taken into account.

For *close protection*, p.v.c. cable ratings can be multiplied by 1.33. Most cartridge fuses and circuit-breakers give close protection so that under these conditions 2.5 mm² size cable would be permitted to carry 21 × 1.33, i.e. 28 A instead of 15 A as shown in the table. Cable selection is so important that another example is merited.

A 240-V electrically heated shower is rated at 7 kW. What is the minimum size of cable if the length of run is 20 m and protection is given by a miniature circuit-breaker?

It is first necessary to obtain the current, from the power equation, which is equivalent to the 7 kW load.

Power (watts) = Voltage (volts) × Current (amperes)
In symbol form $\quad P = VI$
By transposition $\quad I = \dfrac{P}{V}$

$$= \frac{7000}{240} = 29.2 \text{ A}$$

Since a miniature circuit-breaker permits close protection, 4 mm² cable is allowed to carry

27 × 1.33 = 36 A

The cable size must now be tested for voltage drop. As with the previous example

$$\text{Volt drop} = \frac{\text{mV/A/m} \times I \times l}{1000}$$

$$= \frac{10 \times 29.2 \times 20}{1000} = 5.84 \text{ V}$$

Since this is within the 6-volt limit, selected size of cable to fulfil all conditions is 4 mm².

What other cable-insulating materials are used?

Butyl rubber does not suffer from the possibility of conductor decentralization as previously mentioned. The cable may generally be continuously rated at 85°C and contains a higher overload temperature. An additional appropriate heat-resisting fibre-lapping increases the continuous operating temperature to a maximum of 100°C. Butyl rubber core insulation when combined with niplas sheathing for flexible cords complies with BS 6500/1969.

E.P. rubber — a newer synthetic material — stands for *ethylene propylene rubber*, is in accordance with BS 6899 and is one of a class of materials known as an elastomer. While the physical and electrical properties are similar to butyl rubber, this new polymer has improved resistance against the effects of water; and the long-term ageing properties show even improved performance over the higher standard of butyl rubber. Referring to the other end of the scale, e.p. rubber retains a flexibility even at temperatures as low as −70°C.

The most heat-resistant is *silicone rubber*, which also withstands temperatures as low as −75°C. It is also resistant to a wide variety of chemicals and in general will resist the attack of oxidizing agents, weak acids, salts and vegetable oils. One reason for the ability to withstand these temperature extremes is that the material is not thermoplastic and thus retains good dielectric quality and elasticity at sustained high working temperatures.

What types of flexible cords are generally available?

(1) (*a*) Parallel twin, (*b*) twisted twin, (*c*) flat-twin sheathed.
(2) Circular 3-core or more, made up with overall braiding, care being taken against abrasion and that the maximum temperature as specified for the insulation is not exceeded.
(3) Single-core, twisted-twin and 3-core insulated with glass fibre, provided that these types are used only in dry situations for light fittings or for other applications where the cord is not subject to undue flexing.

Where the cords are exposed to a risk of mechanical damage, they must, as a minimum, be effectively sheathed. Armouring provides additional protection, one type of which may be seen in Fig. 7.

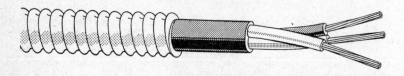

Fig. 7   Three-core p.v.c. cable (*BICC*)

What is double-pole fusing and its possible dangers?

Fig. 8 shows the arrangement, together with a double-pole switch. This is probably a relic from direct current (d.c.) supplies and may still be seen in older premises. A dangerous situation can arise

(*a*) since, with uneven fusing, the neutral fuse only could rupture, leaving the live fuse intact, or

(*b*) by pulling out the neutral fuse and leaving the fuse in the live side.

In both cases, although lights and appliances would not operate, they are alive to earth because neutral conductors are at earth potential. The position is somewhat similar to the situation depicted in Fig. 5.

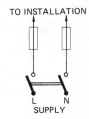

Fig. 8 Double-pole fusing

The lethal character may be emphasised from a DPCP commentary on electrical fatalities:

The deceased was changing a lighting fitting but had only removed the lighting circuit neutral fuse. The live fuse was still in position and the circuit was live. He received a fatal shock when he drew live wires through an isolated section of conduit while he was also in contact with earth.

Installation technology often refers to the term "discrimination". What does it mean and what is its significance?

As we shall see, the term is open to a wide interpretation. In essence, it means that, in the event of a heavy overload or short-circuit, only its local fuse or circuit-breaker will operate. In house wiring for example, if such a fault occurs on any lighting circuit, the fuse nearest to the fault will 'blow' or the miniature circuit-breaker trip. In this way, only that particular circuit will be affected and so avoid plunging the whole house in darkness.

An ignorant householder may rewire the fuse-holder with a fuse-wire of too thick a cross-sectional area (in the mistaken notion that the thicker wire will be safer!) An overload or short-circuit fault will by-pass this heavy local fuse-wire and could result in the Electricity Board's main fuse rupturing and so cut off the entire supply from the premises.

12   *Electrical Installation*

The term discrimination has an even broader significance than the example just mentioned. Large industrial plants and commercial units possess quite sophisticated forms of protection. There may be a main fuseboard, sub-main fuseboard, and local fuseboards for final sub-circuits (Fig. 9). Planning of a 3:1 fuse ratio or circuit-breaker ratio would give good discrimination and follow the sound installation principle that, in the event of a fault in the final sub-circuit, other circuits will not be affected.

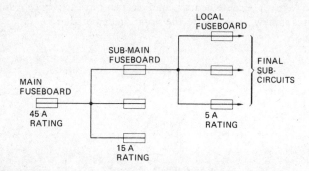

Fig. 9   Diagram of a 3:1 fuse rating giving good discrimination

A further point to be considered is that fuses, where fitted, throughout the system should be of the same type, i.e. rewirable should not be mixed with high breaking capacity (h.b.c.) types.

Clearly domestic bell and mains wiring in close proximity constitute a potentially dangerous situation. Can more specific voltage ranges be stated?

Bell wiring normally comes under extra-low voltage (e.l.v.), whilst installation mains voltage comes within low voltage (l.v.).

E.L.V.   a.c. up to 50 V between line and neutral or between line and earth.
d.c. up to 120 V between line and neutral or between line and earth.
L.V.   a.c. above e.l.v., up to 1000 V between line and neutral.
above e.l.v., up to 600 V between line and earth.
d.c. above e.l.v., up to 1500 V between line and neutral.
above e.l.v., up to 900 V between line and earth.

As a memory aid, these values are illustrated in Fig. 10.

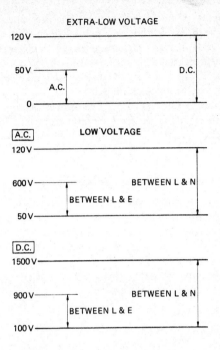

Fig. 10 E.L.V. and L.V. voltages

State some terms which have been renamed in the 15th Edition of the IEE Wiring Regulations (now called "Regulations for Electrical Installations")

14th Edition
Earth-continuity conductor.
Final sub-circuit.
Current-operated circuit-breaker.
Excess current.

15th Edition
Circuit protective conductor.
Final circuit.
Residual current circuit-breaker.
Over current (now separated into protection against (i) overload current, (ii) short-circuit current).

Following the previous question, what new terms have been introduced?

EXPOSED CONDUCTIVE PARTS. Refers to the earth path of an installation which may become alive under earth fault conditions.
EXTRANEOUS CONDUCTIVE PARTS. Parts of earthed sections which are not included in the installation proper such as bonded gas pipes, radiators, and structural metalwork.

LIVE PART. Conductor or conductive part. The neutral conductor comes under this heading and thus becomes a live conductor.
PROTECTIVE CONDUCTOR. General term which covers bonding conductor, circuit-protected conductor, and earthing conductor.
DIRECT CONTACT. Contact with live parts which may result in an electric shock. This also applies to animals.
INDIRECT CONTACT. Contact of persons or livestock with exposed conductive parts (earthed) made *live by a fault* and which may result in an electric shock.

How are protective conductor sizes calculated?

This may best be understood by a worked example using 0.45 second as maximum disconnecting time for portable apparatus and 5 seconds as maximum disconnecting time for a fixed appliance.

The size of protective conductor is required for 100 A load controlled by fuses to BS 88.

Maximum earth loop impedance (from Table 4I A2) = 0.45 ohm
Maximum fault current = 240/0.45 = 533.3A

Fuse characteristic time = 0.1 sec
Protective conductor cross-sectional area

$$= \sqrt{\frac{I^2 t}{k}} = \sqrt{\frac{533.3^2 \times 0.1}{115}} = 15.72$$

Nearest protective conductor size = 16 mm²

N.B. The value of k is taken from III Table 54C.
In general, with conductor sizes which are not more than 16 mm², calculations are not necessary as the protective conductor sizes must not be less than the circuit conductor cross-sectional area.

Chapter 2
Cables and Flexibles
(other than mineral-insulated cables)

What metal is most commonly used as a conductor?

The best conductor, i.e. the metal with the least resistance, is silver but is rarely used mainly on account of its cost; it would also be too soft for wiring purposes. Nevertheless it is employed for specialized items such as the tips of make-and-break contacts and certain fuse elements.

Copper is generally employed as the conductor because it has nearly as high a conductance as silver and is much cheaper. Increased flexibility and greater tensile strength are added advantages.

Are there any practical alternatives to copper as a cable conductor?

Aluminium is without doubt a serious competitor to copper and in fact is preferable where *weight* is the primary consideration. Since its conductivity is some 60 per cent that of copper a larger cross-sectional area (c.s.a.) is required to carry an equivalent current.

The IEE Wiring Regulations do not permit aluminium to be used with a c.s.a. below 16 mm^2 because it is relatively soft and has not the tensile strength of copper.

Due to the nature of the material, care is necessary to prevent overtightening at the terminal connexions. Bimetal action, which can lead to corrosion, must also be avoided; it is caused through direct contact with unplated brass or copper terminals. Similar precautions are necessary at joints between aluminium and copper conductors.

State some precautions for making cable connections

When stripping core insulation, avoid nicking or cutting into the conductor. Correct gauging of the stripped end length is essential for safety: if too long there is a danger of contact with the bare conductor; too short, proper contact will not be made. For single-strand connections, the cable ends should be doubled over if possible so that maximum contact is made with the terminal interior. For safety the flat part of the doubled end should be perpendicular to the terminal screw.

16 Electrical Installation

What is meant by flexible cord?

It may be defined simply as a flexible cable with each core having a cross-sectional area not greater than 4 mm^2.

What sizes of flexible cords are available?

Cross-sectional area (mm^2)	Number and diameter (mm) of strands	Current rating (ampere)
0·5	16/0·20*	3
0·75	24/0·20**	6
1·0	32/0·20	10
1·25	40/0·20	13
1·5	30/0·25	15
2·5	50/0·25	20
4·0	56/0·30	25

*28/0·15
**42/0·15 } for p.v.c. parallel-twin non-sheathed cords. (from IEE, Table 22M.)

From the above table it can be observed that for lighting and light-duty appliances the sizes are 0·5 and 0·75 mm^2. Typical sizes for electric fires are:

1 kW, 0·75 mm^2; 2 kW, 1·0 mm^2; 3 kW, 1·5 mm^2.

What are the maximum weights that may be supported by twin flexible cords?

These are stated in an IEE Table as follows:

mm^2	kg
0·5	2
0·75	3
1·0	5
1·5	5
2·5	5
4·0	5

It is of course preferable to employ straining wires, suitably positioned, which will take the weight away from the cords. Further, when a non-metallic outlet box of thermoplastic material (e.g. p.v.c.) is used for the suspension of a light fitting, care is necessary to ensure that the box temperature does not exceed 60°C.

The mass suspended from the box must not exceed 3·2 kg.

What are the colour-core identification for flexible cords?

Appliances with 3-core flexibles are now identified by:

Live core	Brown
Neutral core	Blue
Earth	Green/Yellow

This requirement now has legal force as it forms part of the Electrical Appliances (Safety) Regulations.

The core colours of other flexible cords as stated by the Commission on Rules for the Approval of Electrical Equipment (CEE) are:

2-core:	Brown	Live
	Blue	Neutral
4-core:	Brown	Live
	Brown	Live
	Brown	Live
	Green/Yellow	Earth
5-core:	Brown	Live
	Brown	Live
	Brown	Live
	Blue	Neutral
	Green/Yellow	Earth

These core colours form a radical change from past practice, so that extra care is essential in checking all new connections. In certain cases, to avoid such errors from occurring, sheathed flexibles can be ordered with sealed moulded-plug termination.

What precautions are necessary for flexible cords in pendants and light fittings?

It will be appreciated that heat-resisting cords are necessary for most connections between the ceiling rose and lampholder where tungsten-filament lamps are to be used, due to the abnormally high temperatures generated by these lamps.

Light fittings, especially if flush-mounted and totally enclosed, require to be provided with heat-resisting insulation suitable for the temperatures likely to be encountered. Where heat-resisting sleeves are used they should be fitted over the individual cores of the flexible in such a way so that the normal insulation of the cores is not relied upon to prevent a *short-circuit* between conductors or an *earth fault*. Similar methods should be employed for accessories and appliances which are subject to such conditions.

For certain types of ceiling light fittings, Fig. 11 illustrates one method of preventing the heat from affecting the main circuit cables.

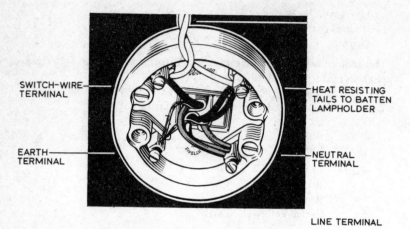

Fig. 11   Ashley Polyblok (*Ashley Accessories Ltd*)

What are the temperature ratings for flexible-cord materials?

The maximum operating temperatures of insulation or sheath of flexible cords are stated in IEE Table 10B as:

	Normal service (°C)	Contained within fittings and not subject to stress (°C)
General-purpose rubber compound	60	75
General-purpose p.v.c.	70	75
Heat-resisting p.v.c.	85	100
E.P. or butyl rubber	85	100
Silicone rubber	150	200

Where the insulation and sheath are of different materials appropriate temperature limits must be observed for both. *These heat-resisting insulating materials are also used for cable insulation.*

Rubber has been used as an insulant since the dawn of electric wiring; however, above 60°C the insulation may become hard and brittle so that it lacks flexibility.

What flexible-cord sheaths are suitable for contact with oil or petrol?

Where such a risk is present or if a flexible sheath is required which will not support combustion, use must be made of cords with a sheath defined as heat-resisting, oil-resisting and flame retardant (h.o.f.r.) complying with BS 6899. There are two grades of this type of sheathing, (1) GENERAL-PURPOSE which includes polychloroprene (p.c.p.) and (2) HEAVY-DUTY which includes chloro-sulphonated polyethylene (c.s.p.). 'Hypolan' is the trade name of c.s.p. as manufactured by E. I. du pont de Nemours & Co. (Inc.).

C.S.P. sheath gives a good mechanical protection and when enclosing butyl-rubber cores allows the latter to run at even higher temperatures than the figures already given. The combined flexible may also safely be employed in such onerous conditions as exist in laundries.

Chapter 3
Steel-Conduit Wiring Systems and Circuits

What is the conduit wiring method?

The system uses a series of tubes which may be of steel, polyvinyl chloride (p.v.c.), copper, aluminium or alloy. The conduits are linked by boxes — usually circular — for drawing-in the cables after erection. These boxes have tapped lugs enabling the fixing of box lids, fittings or certain switch accessories. Occasionally elbow-pieces are used where conduits turn at right-angles, or tee-pieces where tappings are made into conduits, but many electricians deprecate their use (Fig. 12), even of the inspection type, since they tend to restrict the wiring process. An advantage of conduits is the ease by which they allow fresh cables to be drawn-in at any time.

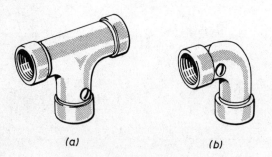

Fig. 12 Inspection conduit fittings
(*a*) Tee-piece (*b*) Elbow

What is the preferred material for conduit?

Steel is generally used because of its great mechanical strength, although aluminium is sometimes adopted for the larger sizes in order to save weight.

What types of steel conduit are available?

Since 1970 conduit sizes have been metricated and are now 16, 20, 25 and 32 mm. The two main types are the light-gauge with plain ends and the heavy-gauge with threaded ends, as made to BS 4568. Further classification as given by the International Electrotechnical Commission (IEC) is dependent upon the type of protection:

Class 1	Light protection both inside and outside	Example	Priming paint
Class 2	Medium protection both inside and outside	Example	Stoved enamel; air-drying paint
Class 3	Medium-heavy protection. Inside as *Class 2*, outside as *Class 4*	Example	Stoved enamel inside; sherardized outside
Class 4	Heavy protection both inside and outside	Example	Hot-dip zinc coating; sherardized

Hot-dip zinc coating is generally known as 'galvanized', contrasting with the zinc-impregnated sherardized finish. Conduit longitudinal seams are invariably welded; the heavy conduits are completely solid-drawn where flameproof work is desired to make the whole system explosion-proof.

What are the available capacities in steel conduits?

Conduit capacities for 600/1 000-V p.v.c. cables are shown on page 22.

Note: The maximum numbers of cables in the table relate to conduit runs incorporating not more than two $90°$ bends or the equivalent. Where runs include additional bends, sets or other restrictions, the numbers must be appropriately reduced. This applies particularly where solid-conductor cables are to be drawn-in.

What is meant by space factor and how does it apply to conduits?

Space factor is the ratio of the total cross-sectional area of the cables to the cross-sectional area of the conduit housing the cables; it is expressed as a percentage. The stated maximum value for conduits is 40 per cent.

Cable size (p.v.c. cables)			Conduit size and cable								
			16 mm		20 mm		25 mm		32 mm		
Nominal conductor size (mm²)	Number and diameter of wires (No./mm)	Nominal overall diameter (mm)	Metric light	Metric heavy	Metric light	Metric heavy	Metric light	Metric heavy	Metric light	Metric heavy	
			Maximum number of cables								
1·0	1/1·13	2·9	8	7	13	12	22	19	38	35	
1·5	1/1·38	3·1	7	6	12	10	19	17	33	31	
2·5	1/1·78	3·5	5	4	9	8	15	13	26	24	
2·5	7/0·67	3·8	4	4	8	7	13	11	22	20	
4	7/0·85	4·3	3	3	6	5	10	9	17	16	
6	7/1·04	4·9	3	2	5	4	7	7	13	12	
10	7/1·35	6·2	—	—	3	2	4	4	8	7	
16	7/1·70	7·3	2	—	—	—	3	3	6	5	
25	7/2·14	9·0	—	—	—	—	2	2	4	3	
35	19/1·53	10·3	—	—	—	—	—	—	3	2	
50	19/1·78	12·0	—	—	—	—	—	—	2	2	

(*Extracted from IEE Table B.5M — Capacities of steel conduits*)

Maximum capacities of steel conduits (BS 4568, Part 1) for the simultaneous drawing-in of single-core p.v.c. cables (BS 6004)

What size of conduit is required for four 2·5 mm² and four 6 mm² p.v.c. cables?

Since the space factor is 40 per cent:

$$\frac{40}{100} = \frac{\text{c.s.a. cables}}{\text{c.s.a. conduit}}$$

$$\text{c.s.a. conduit} = \frac{100 \times \text{c.s.a. cables}}{40} \text{ mm}^2$$

$$= \frac{100}{40}(4 \times 0{\cdot}7854 \times 3{\cdot}5^2 + 4 \times 0{\cdot}7854 \times 4{\cdot}9^2)$$

By reference to Table (p.22) and given circle area = $0{\cdot}7854 \times \text{diameter}^2$, therefore:

$$0{\cdot}7854 \text{ conduit diameter} = 10 \times 0{\cdot}7854 (3{\cdot}5^2 + 4{\cdot}9^2) \text{ mm}$$

Conduit diameter = $\sqrt{362{\cdot}6}$ = 19·1 mm

Nearest conduit size = 20 mm.

How are boxes fitted to steel conduits?

Light-gauge conduits, which are a push fit, are held by a simple lug-grip. Fig. 13 shows how they are fitted to an elbow, but the same principle

Fig. 13   Lug-grip elbow

is adopted for tees and all lug-grip boxes. In order to achieve continuity a band of some 13 mm width at the tube-end must be cleanly filed prior to connection.

Heavy-gauge conduits may be threaded into spouted ends of standard circular boxes. In an almost perfect form of connection there are no exposed threads; surplus threads are a certain sign of bad workmanship as they result in floppy fits which in turn can lead to corrosion and poor continuity.

One form of adaptable box may be seen in Fig. 14; in order to avoid the labour of drilling these boxes are mostly of the 'Knockout' type (Fig. 15). Of the two methods of conduit attachment, the

24   Electrical Installation

Fig. 14   Conduit connection to adaptable box (*Simplex*)

hexagonal smooth-bore bush is preferable although slightly dearer: it allows more wiring space and is definitely a stronger form of connection.

What preparation is necessary?

Conduits are usually made up in random lengths of 3—4 m — 4 m preferred — and supplied in bundles of some 152 m, often simply held

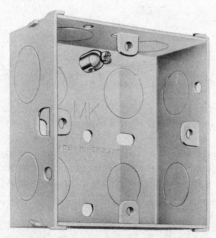

Fig. 15   Standard 'Knockout' box
(*MK Electric Ltd*)

together by string. Thoughtful manufacturers include a complete wrapping of prepared brown paper minimizing the damage to the enamel during transit from works to the site. In addition the electrical craftsman should get into the habit of giving a quick look through each conduit, before use, in order to make certain that there are no obstructions. If any pebbles or similar debris become lodged it may be costly to replace the conduit when the cables are being pulled through, since this operation often occurs towards the finishing stage of the building.

Light-gauge conduit ends are simply prepared by filing a cleaning band, the width governed by the lug-grip fitting. The internal ends are then reamered to avoid the possibility of damage to any of the cables which will be contained in the conduits. Raw sharp edges as left by the hacksaw cut, unless properly treated in this manner, can easily play havoc with the vital cable insulation.

For the heavier gauge, threading is necessary; on sites the threads are cut by means of hand-operated stocks and dies which should not be rotated continuously. Frequent sharp swings in an anti-clockwise direction is a common method adopted for clearing the tube cuttings (swarf). If swarf is allowed to remain it may lodge between the cutting teeth and result in stripped threads — making for floppiness and poor continuity. Regular washing out of the dies with paraffin prolongs their life and helps to ensure clean and perfect threads. Some makers supply tins of special cutting compound, although they are rarely employed. To avoid corrosion lubricants must be of a non-acidic nature and oil or plain cooking tallow are often employed.

How should the ends be reamered?

The electrical worker often inserts the nose of cutting pliers into the tube and then with some pressure rotates the tool to and fro, i.e. in a clockwise and anti-clockwise direction. The correct method is to use a purpose-made reamer of which there are two types: one is hand-held while the other has a square end for fitting into the ordinary carpenter's brace. The sharp cutting edges of the latter permit the tool, in addition, to be employed for opening out holes in metal boxes, say from 16 to 25 mm. However, these are special tools, and so it is common practice to see the operative work with a round file for clearing the burrs without any apparent ill-effect.

How is a conduit bent?

Although light-gauge conduit is often regarded as not amenable to bending, old stagers have been known to make some beautiful sets (as the bends are often called) by drawing the tubing carefully across the knee-cap. With the bending machine, consisting of a stand containing appropriate formers, guides and a lever, there is certainly no problem in setting the light-gauge brazed type. Partly due to this ease in bending this conduit has been specified by local authorities and successfully fitted into many thousands of newly-built multi-storey flats. The portable bending machine is also almost invariably seen as part of the electrical contractor's equipment on building and factory installations utilizing heavy-gauge tubing. Fig. 16 shows the essential parts of a hand-operated bender, which may also be adapted for the Baco aluminium conduit system.

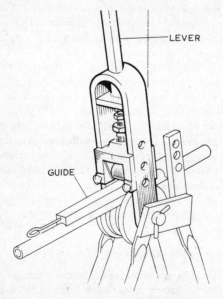

Fig. 16   Hilmor portable bending machine

The worker on the site supplements the bending machine by use of the familiar wooden bending block.

What is a bending block, and how should it be used?

A good block is made from ash, free from knots, of section 100 by 80 mm and some 1·5 m long. The hole for inserting the conduit should be about 150 mm from the top. Diagonally opposite ends of the hole must be chamfered to avoid any flattening of the conduit. Bending takes place by easy and gradual pressure; the block should be tilted forward and the conduit continually moved through the hole as the set is being made. Bends are recommended to be fashioned with as wide a sweep as possible, and in any case the inner radius of the bend must not be less than 2½ times the outside diameter of the conduit. A purpose-made metal hand bender can be purchased; although it saves the labour of making a wood block and has a long life, care in use is required as the conduit may give a nasty jerk when pressure is applied. Fig. 17 indicates some of the names given to the various types of sets.

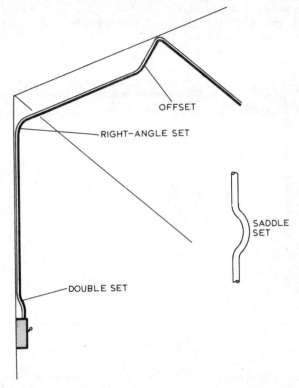

Fig. 17   Some common sets

Where conduit has to follow awkwardly-shaped building contours at a high level, a template marked in chalk on the ground or on a board often saves a lot of arduous climbing of ladders. Tight sets can often be made by bending block if the next larger size conduit — 'a sleeve' — is slipped over the tube to be set.

Since the IEE Wiring Regulations do not permit inspection fittings or boxes to be buried, how can conduits when laid in solid floors be made to form a rewirable system?

For work in pot or concrete-type floors as cast on site, simple boxes without spouted entries but with up to four back-outlet clearance holes avoid the need for inspection outlets. The box outlets serve the dual purpose of drawing-in the cables and the fixing of light fittings (Fig. 18).

Different methods of looping-in may be necessary where specifications require solid prefabricated floor beams. In these cases planning of the conduit runs with the building contractor, structural engineers and other interested parties, may be necessary. An alternative method is to wire with mineral-insulated cables (referred to later) sunk into the

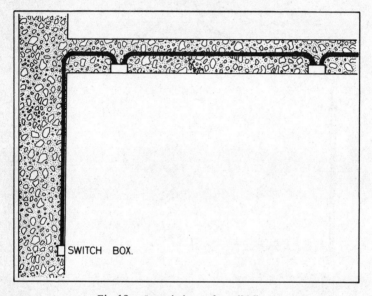

Fig. 18   Loop-in boxes for solid floors

plaster-finish thicknesses and connected to special shallow boxes; another is the use of purpose-made sheet-steel hollow skirtings.

What is the best way to tighten bushes in loop-in boxes as fitted for concealed work?

The nose of the electrician's pliers, as frequently employed, is certainly *not* the best means of tightening bushes and ensuring the minimum electrical resistance between the box and conduit. Examination will also reveal mutilation marks scored into the bushes.

Manipulation by the correct size of box spanner (Fig. 19 (a)) makes for a really tight fit. Where space is so restricted that it is not even possible to insert the box spanner, resort should be made to the ingenious tool as illustrated in Fig. 19 (b).

Steel-Conduit Wiring Systems and Circuits 29

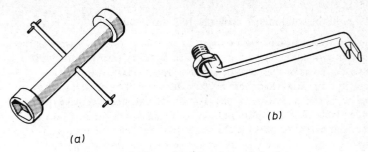

Fig. 19   Spanners for tightening male bushes

How can concealed conduits be picked up for work on the surface?

This always poses a problem and the writer has seen some nasty cutting into finished building surfaces to achieve this objective. One solution is shown by the use of the backless box (Fig. 20), which is specially made for continuing conduit runs on to the surface. There is, however, no reason why appropriate standard boxes should not be adapted for this purpose: care is required in marking out and all holes in the boxes need to be bushed.

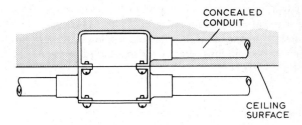

Fig. 20   A method of picking up concealed conduits

What are meant by flameproof installations?

The wiring to petrol pumps is one example. To cover all conditions makes the subject a complicated one. Briefly, flameproof normally refers to safeguarding against (1) the resultant ignition of explosive gases due to the formation of electric sparks (it must be borne in mind that sparking occurs whenever a lamp is switched on or off) and (2) the ingress of such explosive gases into the conduit system. Welded conduits are not permitted — they must be solid drawn; it is compulsory that all boxes, switchgear, etc. which are of the heavy-duty type, must bear the flameproof emblem, which takes the form of the outline of a crown containing the letters FLP, to show that they have been certified by the appropriate authority.

How should the solid metric cables be fed into conduits?

The metric cables of half-hard copper are obviously not as flexible as the earlier standard counterparts, so any sharp doubling over for entry into inspection elbows or tees is to be avoided. Irrespective of inspection fittings, it is essential that these newer cables must be handled with great care to avoid sharp bends or kinks. The careful worker may have drawn the older cables into conduits straight from cable reels mounted on a spindle, and wherever possible this method should also be adopted for the metric cables. Another point to note is that the insulation of an unstranded cable is liable to slip back from the end of the conductor during the drawing-in procedure. It is therefore important to make certain that the insulation and conductor are firmly gripped together so that they cannot slip one upon the other, and for this reason some form of binding may be necessary. In certain situations it may be possible to cover the joint between the cable and draw-wire with black insulation sticky-tape, running the latter on to the insulation. After completion the joint should be sprinkled with french chalk for ease of pulling through. For stripping the insulation, it is advocated that a stripping tool should be used, as a knife may lead to cuts or nicks in the conductors.

IEE Tables B.5M and B.6M indicate the *maximum* number of cables which may be drawn into a particular size of conduit. Experience suggests that it will often be found necessary to reduce the number of single-wire cables as shown in the tables. Due allowance must also be made for restrictions in runs, such as the number of bends and the distance between them.

What further precautions may be necessary?

(1) For vertical runs exceeding 5 m, clamping of the cables is essential in order to take any strain off the conductors and insulation.

(2) Live conductors must not be run by themselves in metal conduits, the reason being that under this condition eddy currents will be set up in the conduits, caused by the changing magnetic field associated with the live conductor 'cutting' the metal conduit and resulting in a temperature rise.

With a current of 16 A passed through an insulated conductor, run through a length of 20 mm conduit, an ammeter showed a current of as much as 2·5 A *passing through the conduit*. Live and neutral conductors when run together effectively nullify the effect of these magnetic fields.

(3) Segregation. *Extra-low voltage*: not exceeding 50 V a.c. or 120 V d.c. (e.l.v.); and *low voltage* (exceeding extra-low voltage but not above 250 V, whether between conductors or between any conductor and earth) must not be contained in the same conduit. This means, as

an example, that normal bell wires as fed from the household bell transformer are not allowed to be run together with cables supplied by the mains, an exception being when the insulation of the extra-low-voltage cables complies with the requirements of low-voltage cables. Cables fed from a mains supply system at low voltage must in no circumstances be drawn together with fire-alarm circuits in the same conduit.

(4) There is a great danger of floors being weakened by conduits being laid in floor joists, so the depth of notches should not exceed that of the conduit diameter. Maximum strain occurs when the cuts are made across the centre of a room, and joists should therefore be slotted close to the bearings.

Light points as wired in conduit may be wired on the 'loop-in' system. How can this method be employed for two light points, one individually switched and the other by 2-way control, the latter to be switched *on* or *off* from two positions?

The term 'loop-in' is derived from the arrangement shown in Fig. 21, where it can be seen that successive light points and switches are fed from cables connected or looped into the nearest light and switch. This method contrasts to the early days of wiring when, instead of looping in and out of the terminals, feeds were obtained by actually jointing into live and neutral cables.

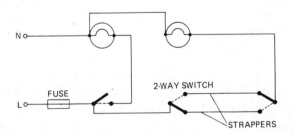

Fig. 21 Loop-in connections

It will be noted that the 2-way switches are of a three-terminal type with a common terminal enabling the current to flow through alternate strapping wires for successive switch positions.

There is also a safety feature since, with the switches in the *off* position, there is no live connection at the ceiling outlet. This contrasts to the 3-plate method as necessitated by the use of p.v.c. twin and earth sheathed cables.

Referring to 2-way switch control, what is a simple method of maintaining the light ON irrespective of each switch position?

The one-way switch (Fig. 22) will keep the lamp illuminated irrespective of the position of each of the 2-way switches. By tracing out the circuit it will be noticed that the one-way switch when on, in effect, shorts out the strapping wires.

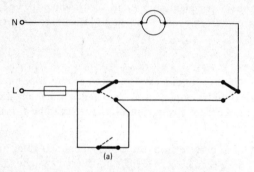

Fig. 22 Switch (a) maintains light on

The term 'borrowed neutrals' is sometimes used in reference to conduit wiring. What are its features?

Fig. 23 depicts the arrangement. While the lights operate individually in the normal manner, the neutral is common to both circuits. This method of wiring may effect some saving in cable but it infringes safety regulations, which stipulate that each circuit is to be electrically separate.

How does the double-pole changeover switch operate? Give an example of its use. Can this type of switch be fitted for any other purpose?

Before answering the question, from a scrutiny of the two previous examples it will be understood that the simple 2-way switch is in reality a changeover switch of the single-pole type. Reference to Fig. 24 (*a*) shows how the double-pole variety, which is also hand operated, enables supply to be made to two completely independent circuits — circuit I or circuit II — but they cannot be put on together.

One important usage for the double-pole changeover switch comes within the area of restricted lighting. The major or normal illumination can be supplied with the switchblades at position I, while at II a secondary circuit gives minimum lighting which might be necessary as a safeguard against vandalism or break-ins.

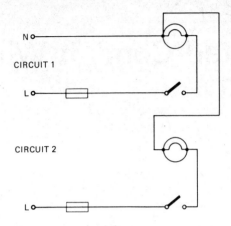

Fig. 23 Borrowed neutral

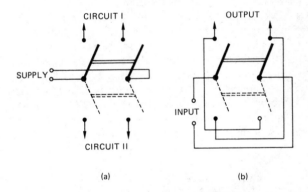

Fig. 24 Double changeover switch connections

With alteration in connections (Fig. 24 (*b*)), the double-pole changeover switch, if fed by direct current (d.c.), produces a reversal in polarity of the output from switch position I to position II and vice versa. Continuous movement has the effect of producing a primitive alternating current (a.c.).

Chapter 4
Insulated-Conduit Wiring Systems

What are insulated conduits?

They are generally made of a p.v.c. material; being supplied in long straight lengths they follow the lines of metal conduits as described in Chapter 3. However, there are a number of important variations which make for a possible wider scope of application.

Black, super-high-impact, heavy-gauge insulated conduit will withstand an impact of approximately 3 kg falling through a distance of about 2·4 m, and is usually supplied, as are also heavy-gauge types, with plain ends, although threaded ends may be obtained. There are corresponding light-gauge insulated conduits which may be coloured white.

Whilst the above are made in straight 3-m lengths, heavy- and light-gauge *flexible* tubing are available in lengths as long as 45 m and have many obvious installation applications. Corrugated tubing also has the flexible qualities, but draw-wires should be left in to avoid cable-ends

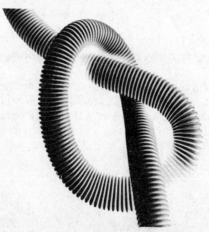

Fig. 25 Polypropylene flexible conduit

catching against the ribbed sides. Polypropylene flexible conduit (Fig. 25) is specially applicable for connecting to motors and switch-panels and may, in addition, withstand flexing and vibration almost indefinitely.

Oval conduit, white, standard 3-m lengths may usefully be employed for switch-drops in order to minimize wall chasing. Oval or round adaptors or couplers are readily available. An interesting version is the oval *slotted* capping. Rectangular channelling, 'D' sections (open or closed), are further indications of the versatility of these insulated extrusions.

Since major installations in Britain use steel conduits, why should plastic types be considered?

Although p.v.c. plastic tubing has been used increasingly since the end of World War 2, there is no doubt that this wiring system still suffers from widespread ignorance and prejudice. Designers who specify its use have appreciated the potential value of conduit manufactured of a material which has many of the advantages of steel conduit (e.g. the rewirable qualities) with little of its drawbacks, and engineers have long been aware of the paradox of first insulating wires and then surrounding them with a conducting envelope of steel conduit. It is with increasing knowledge and experience of plastics that the production of a suitable alternative has become possible. The wiring of such building schemes as the skyscraper London Barbican buildings and a 21-storey block in Paddington are among many schemes that are entirely protected by insulated conduits, whilst certain European countries have a total ban against steel conduits. Holland employs nothing but plastic piping; Italy and France use them considerably without any apparent ill-effects. One thing is certain, there will be a considerable saving over traditional conduit work: taking the cost of conduit alone, the following table indicates the price differential ruling, which is likely to increase:

Price comparison between steel and plastic conduit

Heavy-gauge steel-screwed conduit		Heavy-gauge p.v.c. conduit
Black enamel	Galvanized	
20 mm, 100%	150%	80%
25 mm, 100%	150%	80%

Further economies can be made in the many situations where light-gauge p.v.c. conduit may be substituted. There is also a saving — often considerable — in labour; different techniques are necessary but these skills are soon acquired by competent tradesmen.

How are the conduits fitted to boxes?

Heavy-gauge types may have the ends threaded for turning into spouted boxes, which can be obtained with certain refinements as shown in Fig. 26 In fact, for surface operation the system in appearance may look remarkably like metal-conduit work, threads being cut by a clean and

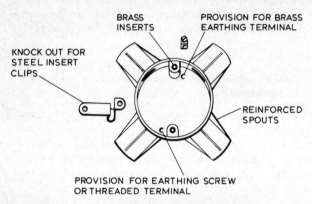

Fig. 26   Threaded plastic 4-way box

sharp set of stocks and dies which is rotated with the arms of the stocks removed. However, this method is not recommended as the depth of threading produces definite weaknesses and can easily lead to conduit fracturing at these points.

Adhesives which are available for p.v.c. conduit do not suffer from any of these disadvantages. The adhesive used for permanently joining p.v.c. fittings is a chemical solution which partly dissolves the surface of the material for a short time and results in a permanent weld between the two surfaces. Tube and fittings must be perfectly clean and free from grease; the cement is applied to both surfaces and the tube rotated within the accessory to ensure complete coverage and within a few minutes a secure joint is made, becoming solid and watertight after a few hours. A Bostik preparation can be used as an alternative and is perhaps preferable in many cases: although also providing a watertight joint it never sets hard but remains permanently tacky to allow for expansion or alteration. It should be borne in mind that the coefficient of thermal expansion of p.v.c. tubing is approximately equal to an expansion of 6·5 mm in a 4-m length for a temperature rise of $45°C$, so that allowance may have to be made for this effect.

Insulated conduits are often telescopic, i.e. each size fits over the next smaller. Small off-cuts of a larger size can thus be used as couplers and, with mastic cement, are ideal in allowing for thermal expansion.

Knockout boxes (Fig. 15) for accepting switches to BS 1299 or 13-A socket outlets, and circular type (Fig. 27) with centres

Insulated-Conduit Wiring Systems 37

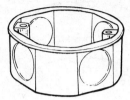

Fig. 27   Circular plastic box (may be fitted with strengthening lugs)

to BS 4568, require clip-on spout adaptors (Fig. 28). One type of universal box has knockout holes provided for 16, 20 or 25 mm conduits. The conduits are held by an ingenious clamp producing a solid mechanical lock, similar in some respects to a bayonet-cap lamplock. This form of connection also allows for expansion, which may

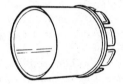

Fig. 28   Clip-on spout adaptor

have to be taken into account in situations subject to rapid temperature changes; also special expansion couplers are available for situations where a great temperature differential is likely to occur.

Are any special steps necessary for fixing?

There is no difficulty in obtaining standard clips or saddles. For surface work it is desirable not to increase the distance between fixings beyond one metre. When fitting circular boxes to ceilings two screws with washers should be used, otherwise the fixing of lug types are advised, the latter being advisable for the suspension of light fittings. IEE Regulations stipulate that where a non-metallic outlet box of thermoplastic material (e.g. p.v.c.) is used for such a suspension then care must be taken to ensure that temperature at the box does not exceed 60°C, and that not more than a mass of 2·8 kg be supported from the box, although makers claim that this value can be increased to as much as 13 kg. The Egatube circular box with steel insert clips (Fig. 29) has been specially designed to deal with this heat problem. The small knockouts allow for inserting the clips which also permit heat conduction from enclosed light fittings.

Fig. 29 Egatube box for the suspension of light fittings

How is insulated conduit bent?

The flexible types pose no problem in this respect. The author has seen them used on sites on the Continent for complete tower-flat installations, an obvious advantage in this age of high labour costs being the short time needed for this work. To minimize the possibility of wiring difficulties, draw-wires are left in corrugated or ribbed flexibles, the joints between the cables and the draw-wire being given a smear of petroleum jelly.

Where conduits are of the rigid type, which is common in Britain, certain techniques applicable to this kind of work must be employed. Sets cannot normally be made by means of a wood bending block or even by the bending machine as used for steel conduits or copper tubing. Therefore new methods have to be mastered for the number of different procedures available, one being cold bending carried out with the aid of an appropriately-sized spiral-spring bending core, which is similar to the plumbers' spring used for bending copper pipes. These coil springs are entered into the insulated conduit in order to prevent collapse at the bend. The 'eye' at one end, to which a wire or cord may be attached, enables bends to be made at the centre or any part of a tube; it also provides a means of holding the spring on a hook so that it can be sharply twisted in an anti-clockwise direction, making it easier to withdraw in the case of an acute bend or set. A point to note is that bends should not be straightened with the spring inserted.

One of the characteristics of p.v.c. rigid conduit is the tendency to straighten out after cold bending and so it is necessary to overset to allow for this reversion. How much? Only by practice can the exact amount be ascertained; one of the tricks of the trade, which is recommended for right-angle sets, is to bend about ten degrees more than is required, bend back five degrees so that the bending spring may be released, and then finally bend back the last five degrees. Strong saddling may be required at either side of the bend for surface work. In this connection it may be noted that right-angle bends are listed in manufacturers' catalogues for those who prefer the ready-made fitment.

Insulated-Conduit Wiring Systems   39

Should heat be used when bending p.v.c. conduit?

In cold weather rigid types may show a tendency to become brittle and are even liable to fracture easily. Under these conditions it is a good idea to give a brisk rubbing over with a rag at the position where the bending is to occur. Hot bending is required in certain situations, particularly with the larger sizes, i.e. 25 mm and upwards. As it is necessary to heat the tube a solid rubber cord must be used instead of a spring because the latter would form ridges and corrugations and be very difficult to withdraw. The conduit should be warmed by boiling water over a spirit stove or lamp – the latter may be a butane lamp or the yellow flame of a blow lamp – ensuring that the conduit is heated over a generous length. After about half a minute with a continuous sideways movement of the heat source (holding the conduit with the hands some 0·6 m apart), the conduit will become quite flexible and can then be bent to form the desired shape.

An alternative method, where a direct flame is used, is to insert the insulated conduit into a steel tube and heat along the outside of it, allowing the heat to travel along and around, thus giving an even heat to the whole circumference of the insulated conduit.

For complicated bends a technique adopted for the metal counterpart may be considered, i.e. making the conduit follow measured chalkmarks on the floor. If the bends have an inclination to distort at the radius, it should be stroked out gently but firmly with a rag or glove. Two final points to remember are (*a*) it may be necessary to hold in position for a few moments until the conduit hardens in the ambient temperature, and (*b*) when the rubber cord is inserted, the insulated tube should be slightly stretched during the process of bending.

How is a system earthed?

A separate independent earth-continuity conductor must be run right through the system. This wire must be of minimum cross-sectional area 2·5 mm^2, insulated and coloured green. This additional conductor may be thought of as adding to the cost of insulation; but against this consideration, a sound earth is brought to all positions and there are not the problems of high-resistance joints which so often face the steel-conduit fitter.

Can insulated conduits be prefabricated?

With suitable precautions against damage complete factory-made units are readily available for any particular mass-produced building arrangement. One insulated-conduit method which is amenable for housing schemes is the small-bore type. Being of small dimensions, flexibility is achieved, each bore housing a single insulated conductor and forming a rather

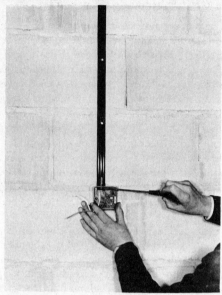

Fig. 30 Small-bore pre-wired insulated conduit
(*Cablecon*)

unique wiring method. The bores which may be twin- or 3-core are webbed together and the tubings are completely rewirable (Fig. 30).

Chapter 5
Trunking and Ducting Systems

What is the difference between trunking and ducting?

Almost identical definitions are given by the IEE Wiring Regulations: 'Duct — an enclosure for the accommodation of cables', and this meaning could, of course, equally be applied to trunking. The two words are frequently used indiscriminately to describe the same thing. 'Duct', however, should be reserved for the *underground* closed passageway or when fixed flush with a finished floor. For trunking, cables are *laid* in place; in ducting they are *drawn* in a somewhat similar manner to the method where conduits are employed.

How do they fit into modern structures?

In present-day steel, concrete and other types of building construction, it is essential that careful thought be exercised in plans and specifications. Buildings cannot be said to be really complete unless there is provision for a normal electrical service and adequate means of expansion of it.

Very many people in the electrical contracting industry have had evidence of the *lack of systematic advance planning* where, in order to remedy the deficiency, it has been found necessary to eventually adopt methods that are expensive, inconvenient and unsightly.

Such remedial work usually results in damage to walls and ceilings; decorations are spoiled and tenants put to inconvenience; walls and floors may have to be cut through, and leads taken to offices or rooms of adjacent tenants, otherwise unnecessary damage may be caused to concealed services in walls. In addition, surfaces are often disfigured through the fixing of casing or tubing. Nor is that all — these difficulties may recur at intervals in the same building.

One method by which all this can be avoided simply depends upon the adequacy of standardized duct or trunking system incorporated in the building structure during construction. In a typical layout of sheet-steel underfloor ducting together with intersection boxes, the lids of the latter are set so as to finish flush with the finished floor. Fig. 31

42   Electrical Installation

Fig. 31   Conduit fed into trunking
(Egatube)

illustrates trunking link-up with conduit. Trunking set into the walls may also follow through to ducting. For work on surface, purpose-made fitments are necessary; an angle or elbow unit is illustrated in Fig. 32. Tees, bends, cross-over pieces, etc. are also available.

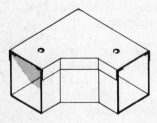

Fig. 32   Trunking-angle unit

Are heat and fire precautions necessary?

Hot air rises, therefore the IEE Wiring Regulations stipulate that in every *vertical* channel or duct or trunking containing conductors, and exceeding 3 m in length, appropriate internal barriers must be incorporated to prevent the air at the top of the channel, duct or trunking from attaining an excessively high temperature.

Further, when ducts or trunking pass through floors, walls, partitions or ceilings, the surrounding gap has to be filled in with cement or similar fire-resisting material to the full thickness of the floor, wall, etc. Any space through which fire or smoke might spread must not be left around the duct or trunking. Where conductors are installed in these types of shafts which pass through floors, etc. suitable internal fire-resisting *barriers* should be installed in order to prevent the spread of fire. Insulated ducts, including sealing compound for joining the various parts, are required to be fire-resistant.

What are the general installation requirements?

Trunking and ducting systems should be designed and installed so as to maintain complete electrical and mechanical continuity, and every part should be provided with devices for establishing an efficient means of bonding and earthing. For sheet-steel types, clean and tight connections are essential. A separate earth-continuity conductor covered in green/yellow insulation must be drawn through ducts constructed of the non-metal fibrous material.

All joints between the various components and parts of the system must be *watertight* under normal conditions of use and especially during any building construction. Metal trunking ducts and accessories need to be protected against corrosion by being either stove-enamelled jet-black, galvanized by the hot process or sherardized (zinc-impregnated) both inside and out, although other equivalent finishes can be accepted. Where enamelled, the paint has to be smooth, continuous, tough and firmly adherent; any galvanized finish also has to adhere firmly to the surface and should be smooth and uniform throughout.

The inside edges of all openings through which cables are to be passed must be smoothly rounded to prevent abrasion, cutting or nicking of the cable insulation. Pin-type racks to act as cable supports are advisable for long runs of horizontal trunking.

On a.c. supply a live conductor must not be allowed to run singly in a metal duct or trunking, but is to be 'bunched' with the neutral in order to prevent eddy currents from being set up in the metal channelling. These *eddy currents* which result in a heat rise are set up by the changing magnetic field which in turn "cuts" the metal trunking. Michael Faraday, the English scientist, in 1831 demonstrated that metal cut or crossed by a magnetic field results in a potential difference. Here the voltage causes small circulating or eddy currents to be set up in the mass of metal.

What are the reasons for the divisions in skirting trunking?

For safety reasons cables containing circuits operating at extra-low and low voltage and supplied directly from a mains supply system must be contained in a separate division from fire-alarm circuits. Low- and medium-circuit cables are also to be separated in this way (Fig. 33) unless their insulation is equivalent.

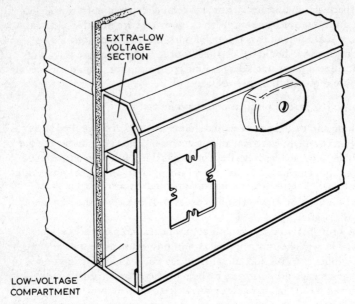

Fig. 33   Trunking outlet facilities

What are the space factors?

As already indicated, space factors are defined as the ratio of the effective cross-sectional areas of cables forming a bunch, to the internal c.s.a. of the trunking. A space factor of 45 per cent is allowed for trunking, compared with 40 per cent for conduits.

The lower space factor for ducting of 35 per cent is due to the possible heat rise in the enclosed duct. The amount of electrical energy responsible for the heat is given by $I^2 Rt$ joules (I = current in amperes, R = resistance in ohms and t = time in seconds). However, the final cable temperature is also dependent upon the rate of dissipation of this heat. The generated heat would, in the ideal case, have been dissipated by self-induced air currents circulated in between and around the cable.

A further factor to be considered is that the total number of cables to be installed in ducts must not hinder the drawing-in of other cables. The 35 per cent space factor relates to runs of ducts incorporating not more than two 90-degree bends or equivalent. On runs having more bends, an appropriate reduction must be made in the number of cables installed. For short straight runs, the space factor of 45 per cent for trunking may be moderately exceeded.

What are rising busbars and why are they used?

They usually consist of bare conductors, of round, square or rectangular section, fixed to insulators and contained in normal sheet-steel trunking, a section of which may be seen in Fig. 34. They may be fed from a busbar chamber, while tappings are taken from the conductors to supply various floors.

Fig. 34 Section of rising main
(*Ottermill*)

Vertical busbars need to take the weight of the conductors; this may be achieved by thrust insulators at the bottom or some form of suitable top supports. Also, due to the change in busbar length as a result of temperature changes, short conductors of flexible material must be interposed on every 30-m length.

What factors are to be considered when selecting busbars?

Copper, with its high conductivity second only to silver, in the cast or unstrained condition is very soft and has low tensile strength, but for many purposes it is essential that it should have these properties increased, and that its limits of elasticity should be such as to give it as high a degree of rigidity as possible under service conditions.

These properties are induced by subjecting it to cold working such as rolling or drawing at ordinary temperatures. Copper is annealed by heating until it gives a dull red glow and then quenching in cold water, removing the oxide scale.

Is aluminium a rival to copper busbars?

Aluminium does not occur naturally in a metallic state, the metal ore being called bauxite; it was first isolated in the 1920s, but difficulty in obtaining sufficiently high temperatures for its separation prevented commercial production for some time. This was overcome by the introduction of the electrical furnace which now produces the metal in a state of great purity (as high a figure as 99·57 purity is said to be generally reached). However, it should be noted that its conductivity is 60 per cent that of copper so that larger sections are required to carry the same current, but set against this disadvantage is the reduction in weight.

What types of aluminium busbars are available?

Aluminium for risers is generally in the form of flat bars arranged singly or multiply; tubular, angle or channel sections give greater mechanical strength, but again greater space is required. Joints are made by welding, brazing, bolting or clamping; to prevent electrolytic action, non-ferrous bolts must be used. Lightness and general ease of installation are distinct advantages here, and the larger radiating surfaces of aluminium busbars result in their running at a much lower temperature than equivalent copper bars.

Where subject to moisture, a liberal coating of bituminous paint is usually sufficient, but aluminium bars can be made proof against corrosion by special processes which employ alternating currents. Aluminium busbars now present a serious rival to copper, the choice depending largely upon the raw-material prices pertaining at the time. In fact, pure aluminium is rarely used; it has been replaced by a magnesium silicide alloy which is designated E91E. There is a slight reduction in conductivity, but this is compensated by the greatly increased mechanical strength.

Where are overhead enclosed busbars employed?

Vertical busbars have been successfully used with appropriate modifications for lateral or horizontal working. In this form they may be sited in factories at truss level, and this has become an arrangement for the distribution of light and heavy loads.

At one time, the normal industrial practice was to drive several machines from one heavy motor; wiring was relatively simple as the power requirements consisted of a few conduit runs only. However, where there is a change from group to individual drives, means must be adopted which make it impossible to touch a live conductor while plugging in. There is thus a natural development towards power points at frequent intervals, which under the older methods would necessitate masses of multi-conduit runs.

In factories and workshops, with changing techniques the layout of machines may be continually subject to modifications. Thus, when plant is altered, new machines installed or works extended, complete flexibility is essential and this can be met efficiently and economically by the overhead busbar system.

The essential point here is that these lateral busbars can be tapped at some 0·6-m intervals to serve individual machines by special fuse boxes (Fig. 35). These have a rigid push-fit, and for safety there are

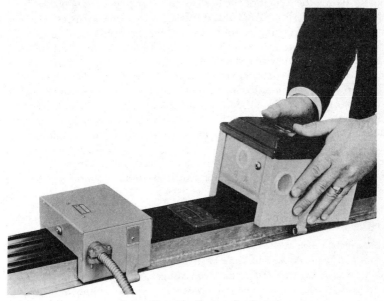

Fig. 35　Fused busbar tapping boxes
(Barduct Ltd)

usually extended guide rods on the sides of the fuse box, enabling positive earthing to be made to the trunking well in advance of the live-phase contactor engaged with those of the busbars; on withdrawal, the sequence is reversed, live conductors being disconnected first. After plugging in, the fuse is securely held by nuts locking on to the retaining rods. There is also a fixed type of fuse box for where it is considered that the machine will be in a permanent or semi-permanent position. In both cases, the fuse boxes are usually triple pole or triple pole and neutral.

What further advantages are claimed for overhead busbars?

Within limits, it is not difficult to favour the lateral busbar scheme which permits speedy re-arrangement of plant, without shutdown of those machines not involved in the change and with the minimum of disturbance. For industries such as the motor-car trade, where there may be a complete annual change to suit the alteration in design, the system shows its superiority over rival methods. It is preferred, in any event, where the shop floor consists of power units close to one another: extensions are readily carried out, and the chasing of floors to sink conduits between fuseboards and machines can be eliminated.

Advantage is taken of modern factory-building construction to simplify fixing; the weight of the busbars and trunking is taken by the horizontal roof cross-members of the frame girders or roof trusses. The run of the busbar trunking on roof trusses through a factory now becomes virtually equivalent to an extended fuseboard, and means that no part of a floor area is more than a short distance from a supply point of ample capacity. It is perhaps not surprising, therefore, that under these conditions power can be installed in a machine shop in advance of the machinery and even before the final plant layout is known. Also, as the system consists of materials which deteriorate very little with age, it is practically indestructible. Should stripping be required, due to alterations or other reasons, an important point to note in this period when materials are expensive is that the scrap or recovery value is high, in marked contrast to wiring in metal or insulated conduits which become almost valueless when stripped.

What ranges are available for busbar trunking?

One standard range in the current ratings, to cater for the heavier loads, is from 150 to 1 500 A; miniature busbars range up to a total load of 100 A. In assessing the load on each run of busbars, the actually working conditions of the machines and apparatus needs to be known in advance. It is normally assumed that they will not be operating all together at full load for the whole of the time, therefore an economy in busbar rating can be made by multiplying the total connected load by a figure of less than 1·0, known as the *Diversity Factor*. One manufacturer advises a diversity factor of 0·4 for a machine shop of individual motorized units, but anything up to 0·8 for a welding load.

Voltage drop for long runs may be reduced by feeding the busbars at a centre point or forming into a ring main. It is common to install p.v.c. armoured cables with compression glands for the busbar supply.

What insulation is used between the busbars?

The earlier types used air for the insulation and this medium is still often employed. However, an insulating busbar sleeving of extruded phenolic material has the advantages of low moisture absorption and great mechanical strength, and is also said to give added protection against the possibility of flashover between phases. The other big advantage in the use of solid insulation is that the overall cross-sectional dimensions of trunking are considerably reduced since the bars may now be fitted closer together. Makers have even designed insulation that is so robust that there is strong mechanical protection — thus under these conditions complete metallic enclosures are rendered unnecessary.

What developments have been made from the metal trunking?

One example is the Atlas Trakline which is made from p.v.c. extrusion (Fig. 36) and is tough and rugged. It is adapted for lighting and small

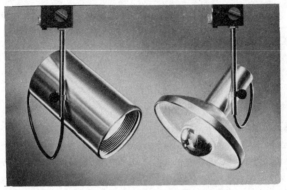

Fig. 36   Atlas Trakline trunking
(*Atlas*)

loads since the busbars which are contained inside the trunking are designed to carry a maximum 7·2 kW at a working voltage of 240 V. Tappings may be obtained by adaptor boxes which are fitted with a 3-A fuse as standard equipment. An earth system is provided by a continuous aluminium conductor which ensures that the earth contact is made before the electrical contact or mechanical locking takes place. The moulded trunking lends itself to supporting fluorescent fittings while the system may be used to feed spotlights, tungsten lamps and small appliances.

What developments have been made with plastic trunking?

Sizes are now equivalent to most of the metal types. In addition to its characteristic features given by its manufacture from high-impact p.v.c., methods of installation may follow those of steel conduit but with certain improvements, e.g. the $90°$ slow bend (Fig. 37) reduces sharp cable bends as demanded by the elbow or mitred accessory.

   Manufactured tees and cross-tee fitments for linking trunking may be purchased, but as the p.v.c. material lends itself to ease in working, many fitments can be made on site.

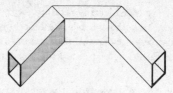

Fig. 37 A 90° slow bend

Gilflex Architectural Trunking incorporates three separate compartments, each with a separate cover. The centre compartment will normally be employed for power supply wiring, with the two outer compartments being used for extra-low voltage circuits such as Post Office telephone, video, computer, intercom, and other similar circuits.

What types are advisable for rewiring?

An examination of reports indicates that some four million homes in Britain are at least 30 years old and have reached, or are reaching, the need for rewiring. Apartment flats with solid floors are not normally amenable to rewiring with p.v.c. twin and earth sheathed cable. It is therefore perhaps not surprising that the use of white p.v.c. minitrunking is on the increase for this purpose.

The white finish is particularly attractive for this purpose and is eminently suitable for surface installation. With its reduced dimensions, minitrunking is fast becoming a wiring system in its own right. For siting above skirtings (Fig. 38) there is no difficulty in looping into and out of plastic outlets for ring circuits; for 90° bends there are internal and external fitments. Ugly clips which inevitably form dirt traps are not required.

25 millimetre square trunking may almost be considered as square conduit! Where fitted in the angle between wall and ceiling, the trunking lends itself easily to form tee junctions for the light points and switches.

Further aesthetic features are provided by the many variations with architrave and cornice profiles.

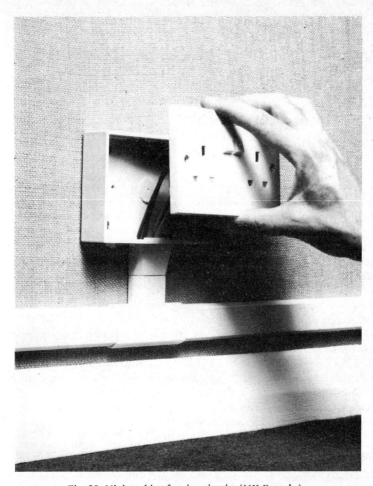

Fig. 38 Minitrunking for ring circuits (*MK Egatube*)

Chapter 6
Mineral-Insulated Cables

The mineral-insulated, metal-sheathed type of cable normally consists of two materials only, namely a highly conductive seamless copper sheath enclosing one or more unstranded copper conductors which are insulated with a tightly-compressed white magnesium oxide (MgO) powder. During manufacture, the sheathing, insulant and conductors are drawn-out after assembly to standard dimensions. Thus the smaller sizes of cables are obtainable in longer lengths (as much as 300 m) than the larger sizes. The advent of the cable known as 'pyro' after Pyrotenax Ltd, the introducers of the system to Britain, has coincided with a need for the reduced, or complete, elimination of the builder's wasteful cutting away. This operation is, in any case, not usually permitted by structural engineers on pre-stressed floor blocks. The original manufacturers have been joined by the giant BICC who are now probably the major producers.

Sometimes abbreviations for the cable are written as m.i.c.c. (mineral-insulated copperclad), m.i.c.s. (mineral-insulated copper sheath) or simply m.i. (mineral-insulated).

How are the cable ends prepared?

Because the mineral insulation is hygroscopic, i.e. it absorbs moisture, sealing of all cable ends is required. Formerly, simple seals consisted of internally-threaded brass pots, tapered to cut threads on the cable sheath in order to form a tight fit.* The stub cap which is pressed on by the crimping tool (Fig. 40) closes the opening after the pot has been filled with a waterproof sealing compound; the crimping tool then securely locks the cap in place. P.V.C. or neoprene sleeves are threaded through holes in the disc to insulate the conductor ends for connecting purposes, and glands allow this to be made to

---

*A modern type of wedge, pot and stub seal (Fig. 39) simplifies the operation. A wedge and stub crimping tool, with the wedge pusher set to the forward position, draws the pot securely over the wedge.

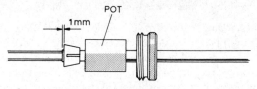

Fig. 39   Wedge-type seal

threaded boxes. Sealing makes the terminations more difficult than with other wiring systems, although with practice the necessary technique is soon acquired.

As part of the gland, the cone-shaped compression ring or 'acorn' is similar in many respects to the ones used for union fitments to small-bore water-heating pipes. It makes a firm — almost watertight — grip on to the cable.

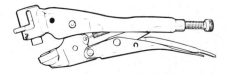

Fig. 40   Wedge and stub crimper
           (*BICC*)

What further precautions are to be taken in making seals?

The operator's hands should be washed before making the seal since absolute cleanliness is essential; metal particles or other foreign bodies will, if present in the seal, soon lead to a breakdown. After fixing on the pot and before filling with compound, the careful worker will carefully examine the pot interior and make sure that any loose powder is knocked away. The use of insufficient compound is a false economy, and filling should take place from one side to avoid the formation of air pockets. After the seal has been completed all surplus compound should be wiped away.

On no account must there be any delay in completing the seal once the cable has been stripped otherwise, if exposed to the air for any length of time, the insulation value will rapidly fall with the absorption of moisture and dampness may be retained permanently in the cable. Puncturing of the sheath, even to the extent of a pin-hole, also results in the absorption of moisture from humid atmospheres.

Operators responsible for installing mineral-insulated cables should make themselves familiar with manufacturers' instructional booklets; conscientious operatives will analyse their own actions and seek improved methods.

What are the working temperature limits of m.i. cables?

One of the outstanding features of m.i. cables is their ability to withstand very high temperatures. As an extreme view, it has been stated that this type of wiring will be unaffected even though the building in which it is installed has burned down. The cable is in fact only limited by the melting point of copper, which is in the region of $1\,000°C$; the melting point of the magnesium-oxide insulant is even higher, being stable up to $2\,800°C$, after which it fuses and forms a glassy substance.

In practice the temperature limits are set by the type of insulation used for the sleeving tails and the type of sealing compound available. Both silicone rubber or p.t.f.e. (polytetrafluorethylene) sleeves may be used up to $135°C$; BICC grey plastic sealing compound is suitable for $80°C$, $105°C$ and $135°C$ seals; the medium plastic compound matches the silicone rubber and p.t.f.e. for use up to $135°C$. Glazing flux with porcelain beads is suitable for exceptional temperature seals as high as $350°C$; for elevated temperatures the standard fibre disc is replaced by the glass-fibre type.

Describe the uses of some of the specialized tools

Stripping off the ends of the copper sheath is a process somewhat akin to peeling an apple, but in this case the stripping is continuous and can be carried out by a pair of side-cutters. The action starts by making a small longitudinal cut at the end to permit a portion of the sheath to be prised up and then tightly gripped by the cutters. By means of a rolling and spiral forward motion, continuous stripping can take place. Some improvement over this method is provided by a kind of home-made sardine-tin opener or purpose-made fork-ended stripping tool with a long looped handle. Another improved tool has a holder which carries a cutting blade centrally pivoted and adjusted by means of a knurled wheel which is then locked with a wing nut. This tool produces a clean circular finish to the stripped-off portion of the sheath; otherwise a 3-wheel cutter is required to make the ring mark and this demands care in use so that the indent is cut to the correct depth. Too little depth results in sheath tear, while too much may cause sheath contact with the conductors; incorrect depth may also make for difficulties when screwing on the pot.

An innovation is the advent of a power-driven tool which strips the sheath at a very fast rate, useful and economical where many lengths of long tails are required. Pots must be squarely threaded on to the cable, possibly by hand, although again care is necessary to avoid cross-threading; the special pot-wrench tool avoids this possible hazard and is certainly a great improvement.

How are the cable conductors identified?

In one respect most m.i. cables suffer in comparison with other types of cables in that the core coverings are not coloured. Multi-cored cables must be checked by a bell-test set or similar method, and then identified by some means, such as coloured p.v.c. adhesive tape available in black, white, red, yellow, blue, green, grey and orange, One of the cable manufacturers has produced an identifiable conductor: prior to assembly a minute groove is cut longitudinally along the length of one conductor, which is then filled with compressed magnesium-oxide powder which leaves a continuous and permanent white line.

What other precautions are necessary?

(1) While the cable can withstand hammering, care must be taken to avoid piercing of the sheath by sharp instruments — any minute hole can produce severe low insulation resistance. When large coils are delivered on the job, the hessian or other protective wrapping should be retained on the cable until it is actually required to be used. Similarly, it is advisable to retain pre-assembled cable units in their boxes right up to the last possible moment; on no account should they be laid on stones with sharp edges. Where the cables are set solidly in concrete the keen electrician will zealously guard them until pouring takes place and the cement mixture is set. Exposed m.i. cables rising from the ground must be protected to a height of at least one metre by metal channelling or bushed conduit.

(2) The regulation for bending states that the minimum internal radius of the bend must not be less than six times the overall diameter of the cable sheath (Fig. 41). The fact that the copper sheath *work*

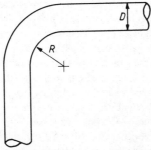

Fig. 41　Internal radius (R) of the bend must not be less than six times the cable diameter (D)

hardens must also be taken into account. Thus for subsequent straightening and re-bending, it may be advisable first to re-soften the copper sheath by bringing it to red heat with a blowlamp or torch.

(3) To prevent the cable spreading fire or smoke through floors, walls, partitions or ceilings, the holes have to be filled in flush to the full thickness of the floor, wall, etc. When the cable is threaded through structural ironwork, the holes must be effectively bushed to avoid abrasion of the sheathing.

(4) Copper is subject to attack through chemical or electrolytic action, and wherever such dangers exist a p.v.c. oversheath is required. To minimize the possibility of corrosion of the seal and gland, the use of a corrosion-resistant hood or shroud as shown in Fig. 42 is required to overcome this problem.

Localized protection against the corrosive effects may be given to short lengths of cable by lapping with p.v.c. adhesive tape. This protection is particularly important in positions where the cable emerges

Fig. 42 P.V.C. covering
(*BICC*)

from a concrete floor, since corrosive elements are likely to accumulate around it. To forestall any possibility of trouble, the tape should be applied (before the floor has been laid) extending to about 150 mm above and below floor level. An alternative protection may be provided by the use of a p.v.c. tube.

The corrosion by electrolytic action is also caused by contact of dissimilar metals which, in effect, forms an elementary primary cell. There may therefore be an actual migration of ions or eating away of one of the metals. It is therefore advisable to minimize any contact with ferrous materials, particularly if dampness is present.

(5) Mineral-insulated cables from 35 mm^2 upwards are only available in the form of single core, and where several of these larger sizes make up a run in parallel, care must be taken to neutralize the *inductive* effects which may be manifested as eddy currents, these being small circulating currents in the copper sheath. These single-core cables should not be spaced apart but must be run in close proximity; where runs comprise three cables, the trefoil (triangular) formation is advisable in order to reduce heat rise and losses.

(6) In the few cases where m.i. cables are fitted as overhead lines it may be necessary to offer protection against the effects of *voltage surges induced by lightning*. This is achieved by surge diverters, connected between the conductors and earth at the incoming supply lines, a form of non-linear resistor, which is practically an insulator at mains voltage and becomes increasingly conductive as the voltage rises. The diverter thus provides an alternative parallel path for any surge voltage.

How does the current-carrying capacity compare with p.v.c. cables?

As may be expected, mineral-cable ratings are higher, as shown by general extracts from IEE tables:

Cross-sectional area (mm^2)	p.v.c. non-armoured single core. Two cables enclosed in conduit or trunking (amperes)	Heavy-duty exposed twin m.i. cables (amperes)
1·0	11	19
1·5	13	24
2·5	18	32
4	24	41
6	31	53
10	42	71
16	56	94
25	73	124

Light-duty or heavy-duty m.i. cables not exposed to touch show an even higher current-carrying capacity for the same cross-sectional area. The reason for the increase over the p.v.c.-cable figures lies in the mineral insulation which can operate at higher temperatures without deterioration.

What methods are adopted for concealed work?

Mineral-insulated cables are easily placed in mixed concrete and similar work, the method depending upon local site conditions. Very often the cables may be laid, without further protection, directly on to the shuttering prior to the pouring of the concrete floors. Although this method is used extensively and shows significant savings in installation costs, care is required to avoid cable damage during the constructional work: gland terminations would be required for connection to conduit-type boxes. Where the cables are run underground they must be buried to a minimum depth of 0·5 m and covered with soil — not coke or ash clinker which are corrosive — and they must also have an overall extruded covering of p.v.c.

In addition the cable lends itself to simple adaptations in the older buildings where a small amount of raking out of mortar or cement permits the pliable cable to be set in between bricks or masonry. Thus, with some careful planning, the cables can be almost completely hidden without the need for cutting into walls or ceilings.

How are mineral-insulated cables run on the surface?

Here cablework is practically the same as that used for the now outmoded lead-covered cables. The slim dimensions give a pleasing appearance.

Instead of buckle clips, fixing is by copper or brass saddles which are held by brass round-head screws. For the smaller sizes hand manipulation is all that is necessary for bends or for snug fits to building profiles; larger cables often need a purpose-made steel bender lined with leather strips to avoid indentations. The cables may require to be finally 'dressed' by a wood block, especially for multi-parallel runs, and cable trays are becoming a common feature for carrying groups across horizontal spans.

How may the cables be made suitable for modern industrialized building methods?

Mineral-insulated cables can be supplied already cut to exact requirements and terminated with glands, seals and earth tails. Each unit is coded for circuit and conductor identification, and also tested at the works for voltage and cable insulation. This is clearly an advanced form of wiring and, while perhaps not suitable for the 'one-off' job, it is ideal for industrialized building work where many similar units are required.

Modern mass-production methods in the building industry demand that construction is designed to a module system. For this reason certain projects such as blocks of flats and offices, schools, hotels and residential estates frequently contain wiring which is repetitive in nature; similarly, certain equipment may entail repetitive internal cabling. Multi-storey blocks may also consist of flats or offices of identical layout which require power and lighting cables of similar lengths.

For these repetitive installations considerable savings in time, and therefore cost, can be made by the use of prefabricated m.i. wiring units, and these can even be made up in a contractor's own workshop. Details are taken from plans or elevation drawings and the cables of standard type or p.v.c. oversheathed are cut to exact dimensions. The units are usually prepared to kit form and packaged accordingly and on arrival at the site can be placed in each dwelling or suite of rooms. In most of these prefabricated systems a trial kit is supplied initially so that any adjustments in cable lengths are corrected for bulk production. Following this stage, at the appropriate building time, the units are placed in position ready to be set into the plaster or cement with the minimum delay to other work.

Do m.i. cables lend themselves to earth concentric wiring?

In this system the metal sheath is used as the return, i.e. it acts as both earth and neutral and, since the sheath can be used externally, m.i. cables lend themselves admirably to this method. There is an obvious gain in replacing two cores by a single core for normal wiring purposes, whilst three-phase and neutral may be supplied by 3-core cables. In addition, as the copper sheath has a much lower resistance than the internal conductor for all sizes up to 10 mm^2, there is a reduction in voltage drop for any given run length. To cut costs even further, suitable backless boxes have been produced.

Earthed concentric wiring is clearly subject to certain fundamental limitations. It is only permitted where the Electricity Board has been specially authorized, by the Department of Energy or the Secretary of State for Scotland, to allow additional connections to be made from the neutral conductors to earth (normally by protective multiple earthing — p.m.e.). Otherwise the supply must be obtained by means of a transformer or converter, so avoiding direct metallic connection to the public supply; alternatively, a private generating plant may be employed.

For earth concentric wiring purposes, the sealing pots are provided with earthed tails crimped to the inside of the pot so that the cable sheath now becomes the combined earth—neutral conductor. At switch and outlet terminations, bonding may require special attention: as the sheath carries the full current, reliance cannot be placed simply on clamping and, in this respect, copper-alloy boxes have an advantage over cast-iron or malleable types.

Can aluminium be used as a conductor for m.i. cables?

Aluminium conductors are permitted for cable sizes of 16 mm^2 and above. However, to achieve economy, *cable with copper conductors and an aluminium sheath* is available for a range of sizes as given in the IEE Wiring Regulations, the cable being covered with a grey p.v.c. oversheath.

Sealing and wiring techniques are similar to the all-copper types, except that particular care must be taken to avoid corrosion by the proximity of dissimilar metals.

Seals so far described appear to involve quite a number of operations. Is there a simpler method?

Work required for stripping cannot be avoided, although it can be minimised by special tools. Time can be saved with the new BICC Pyrotenax Shrink-on Seal (Fig. 43). The method is somewhat revolutionary as the termination requires pre-heating of the outer sheath where the cable terminates.

Heating can be achieved by any proprietary make of gas torch. BICC however have designed a special Pyroshrink torch for this purpose and it ensures that the heat is not only confined to the area required but is also evenly distributed — both during the preheating procedure and that of the final sealing a few seconds later.

Fig. 43 BICC Pyrotenax shrink-on seal

A further feature is that the Pyroshrink torch has a built-in piezo-igniter and, unlike conventional gas torches, has a 'dead-man's trigger' for safety. The handle is shaped for a firm comfortable grip and there is a fold-back stand to steady the torch when not in use.

The termination operation is commenced immediately after stripping. Following pre-heating the cable in the torch exhaust gases, the seal is pushed firmly into place over the cable sheath, after which the seal is heated in the heat chamber until the sealant is observed to flow, and the outer jacket shrinks back fully over the copper sheath and conductors. Finally the required length of insulating sleeving is made to slide over the conductors on to the seal stubs.

A slight variation in method is required for earth-tail seals.

Chapter 7
House and Other Wiring

What cables are fitted in residential premises?

Almost all houses and the smaller blocks of flats are wired with p.v.c.-sheathed cables (Fig. 44); also many shops and offices. Altogether they account for more than half of the national load.

Lighting cables would consist of either the 1·0- or 1·5-mm^2 cables and with this current rating overloading is unlikely to occur.

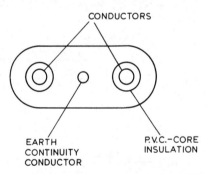

Fig. 44   P.V.C. twin- and earth-sheathed cable

How are lighting circuits wired with the sheathed cables?

For the wiring of new houses and flats containing wood-joist floors, the cables are placed in position before the floors are laid. The layout of the cables is extremely simple, but it is important that the earth-continuity conductor is run right through the installation. In order to comply with the IEE Regulations there must be a proper earth terminal point at each light and switch position, to which the earth-continuity conductor is attached.

Control of a light point by more than two positions requires the use of one or more intermediate switches which must be fitted in the cable run between the 2-way switches. One-way, two-way and intermediate switch control is shown in Fig. 45. The correct form of connections at the 2-way switch nearest to the light point avoids the necessity for any joints behind the switch.

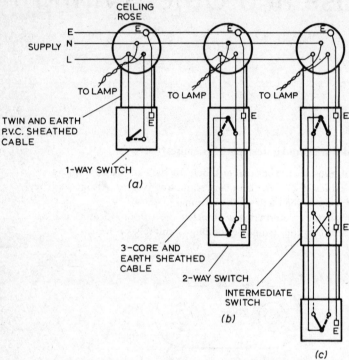

Fig. 45 (a) 1-way, (b) 2-way, and (c) intermediate light control

The 2-way switch is normally employed as a means of controlling light points from two positions. Can it be used for any other purpose?

An examination of the action shows that the 2-way switch is, in fact, as previously stated, a single-pole changeover switch. A simple application permits a bell or buzzer to be sounded (Fig. 46).

When two lamps, individually switched, are required to be controlled by a master switch which will maintain the lamps ON irrespective of the local switch position, and another master to maintain the lamps OFF, what is the basic circuit?

The circuit is set out in Fig. 47. It will be noticed that there is an unconventional use of the double-pole switch for the master ON.

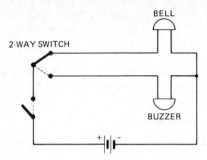

Fig. 46 2-way switch enabling bell or buzzer sounding

An alternative method is the employment of an electromagnetic relay, but its extra complication is hardly merited for this circuit.

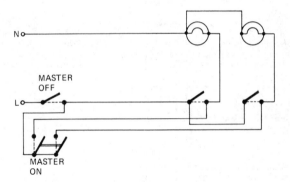

Fig. 47 Master controls

What is the 3-heat switch?

The action of the switch is actually designed for series or parallel circuit connections and may thus be used for specialized lighting circuits. When used in conjunction with heating elements, the four positions allow a flexible control to be obtained, namely: off, low, medium and high. The circuit is shown in Fig. 48 with sheathed wiring.

Suppose each element has a resistance of 60 ohms and is fed by a standard supply of 240 V, then the power obtained by the series – parallel switching may be illustrated by some simple calculations:

Low Two elements in series produce a combined resistance of 120 ohms

$$\text{Power} = \frac{V^2}{R} = \frac{240^2}{120} = 0.48 \; kW$$

Medium One element across the 240-V supply gives a power of
$$\frac{240^2}{60} = 0.96 \, kW$$

High Two elements in parallel results in a power of $1.92 \, kW$, corresponding current values being 2, 4 and 8 A respectively.

How can savings on cable runs be made?

Economy of cable runs can be effected by joint boxes, usually of the brown bakelite type — although for surface work, white boxes are obtainable to match light decorations. Joint-box work shows further advantages, i.e. only one single twin and earth cable is brought out to the light outlet so that a bunch of cables ending at the ceiling rose or batten-holder is avoided. When many cores are exposed at the ceiling roses, extreme care is required for correct connecting up. Apart

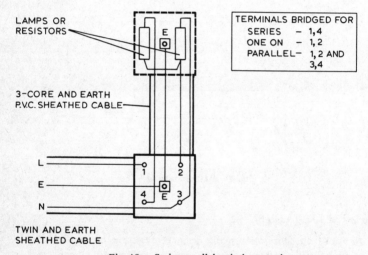

Fig. 48   Series-parallel switch control

from the possibility of connection errors, accuracy in stripping correct lengths is required, otherwise there is the danger of short-circuits, resulting from direct contact between live and neutral conductors. Possible hazards may also occur through contact between the earth wire and a bare live conductor or by direct contact between the neutral and earth. Single-stranded metric cables are especially prone to these dangers as the insulation slips back easily at the cable ends. The joint-box method appears to overcome these difficulties but it is really, in the main, only applicable for use in the older or occupied premises.

Wherever possible a larger size of joint box than the standard 4-terminal type should be employed to allow extra room for stowage of the cable ends.

Porcelain *Scruit* connectors are sometimes used for linking together the earth-continuity conductors. Care is required as the earth wire can too readily slip out of the connectors while the twisting operation is being carried out; thus the channel-type connector is preferable. Six-terminal joint boxes are available and have every advantage — except cost, which limits their use in this highly competitive industry.

A great deal of installation work is concealed and the conscientious worker will carry out his work equally well whether it is hidden or open to inspection. Fixing the joint boxes on wood 'platforms' between joists, whether under floors or in lofts, allows proper runs without any strain on the terminal connections (Fig. 49); those which are simply supported by floppy cables should be avoided at all costs.

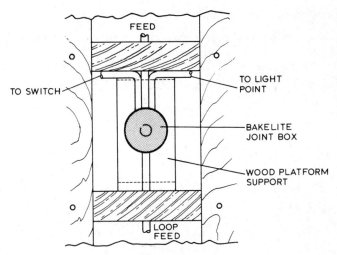

Fig. 49   Joint-box support between joists

What care is required in laying the cables?

Bending requirements are that the minimum internal radius bend must not be less than three times the overall diameter of the cable where the sheath does not exceed 10 mm; four times for diameters between 10 and 25 mm, and six times where exceeding 25 mm.

When buried there must be adequate *protection* against mechanical damage by enclosing in bushed-metal conduits; the oval type is usually preferable. Alternatively metal channelling may be employed.

For surface work *fixings* are often made by means of buckle clips, but recently the plastic pinned clip has come much into vogue. The latter avoids the need for separate pins and clips, but as against this advantage it must be seen that buckle clips can easily be undone for cable-replacement purposes.

Short waste lengths of cable may be cut up and made into effective strap saddles for carcass work in new buildings. The smooth edges of these straps eliminate the possibility of cutting into the cables and even permit fresh ones to be drawn through 'home-made' clips, unclipped cables being easily damaged. On building sites, festoons of wire are all too often seen and these loose cables can easily be wrenched away whilst carrying out other building operations. Another bad practice is that of stringing the cables on cut nails, the sharp edges of which are liable to cut into the insulation.

How should the cables be laid under floors?

When running at right angles to the joists they should be at least 50 mm below the underside of the floorboards. All cables should be drawn in *without twists* through drilled or augered holes in joists. Preferably the holes should be made approximately at the neutral axis (Fig. 50), i.e.

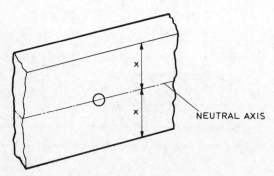

Fig. 50   Holes drilled in neutral axis produce minimum weakness to joists

half-way between the top and the underside of the joists. In this way the cables escape the danger of penetration by nails as the floorboards are fixed and there is no weakening of the flooring; moreover, the holes help to maintain the cables in place.

For hole-making the long, scotch-eye auger provides a useful tool, but a special wood bit fitted into an electric drill is an improvement: it reduces labour and, where an electric supply is available, the arrangement is ideal for rapid work; in addition, a beautiful smooth bore is produced. The small overall dimensions enable the complete assembly to fit between joists for cutting dead straight holes parallel to the floor. As an improvisation the cross-piece can be sawn off the auger and fitted to a heavy-duty electric drill.

Notching of the wood joists is a certain sign of botched work. Apart from making them vulnerable to nails the cables may easily escape from the notches and become pinched between the floorboard and joists.

Loft wiring may not necessitate holes to be drilled in the timber joists since the vertical height of the latter is often only about 100 mm. Where it is necessary to cross they should be run close to the main beams for protection, minimizing the danger of being trodden on.

P.V.C. cables, being thermoplastic, are subject to melting by heat. Such heat can be generated by friction if cables are tugged too rapidly through joist holes, so that care is required as worn or melted insulation represents extreme danger. Petroleum jelly, being chemically inert, does not affect the insulation and may be used on occasion to ease the cables through. In planning runs, one must avoid placing cables above hot-water pipes and certainly not in contact with such pipes.

What methods should be used for wiring with p.v.c.-sheathed cables in inaccessible positions?

Recommended hints come under the heading of 'fishing', for which the flexible nature of the p.v.c.-sheathed cable lends itself admirably. Cavities in older premises, which may require rewiring or some additional points, can often be used to conceal the wiring. One method of fishing out the cables to switch or socket-outlet positions is by means of a 'mouse', made from a bricklayer's white-string chalk line, to which is attached a lead weight, bob or chain. Fig. 51 shows the line being hooked out by a stiff wire hook, the location of the line being aided by means of a small electric torch. Where there are slight obstacles, these may be freed by moving the mouse sharply up and down; in this way the mouse can often find a free space. A draw-wire is made on to the line, and in turn the cables are attached to the draw-wire. Experience plus an intelligent approach minimizes the labour, cutting away and undue damage to decorations.

Horizontal fishing can often be used for wiring across trapless lofts. Luck plays its part but a lot of difficulties can be avoided by the aid of some old unsheathed cable acting as a draw-wire. Contrary to what one might expect no attempt should be made to send the cable towards the other hole. A stiff, galvanized iron wire from this second hole has only

68 Electrical Installation

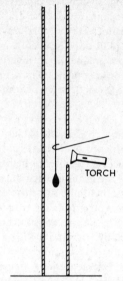

Fig. 51 Fishing cable out of cavity wall

to hook on to any of the wire coils, thus enabling the draw-wire ends to protrude from both holes. Vigilance may be necessary to prevent the stiff wire itself, while travelling forward, from becoming locked with any impediment. This galvanized wire should proceed forward freely and, if it misses any of the coils, one or more twists with a winding motion is usually successful. Cable fishing is an art well worth cultivating for the electrical operative. Hours of frustration can be spent in attempting to fish, either horizontally or vertically, by incorrect methods with consequent loss in labour and temper.

What wiring is suitable for farmyards?

Here electrical installations are subject to severe, onerous conditions. Danger to cables arises from the mineral acids and alkalis present; also steam, ammonia, sulphur fumes and lactic acid all contribute to corrosion of normal wiring. For these reasons plain p.v.c. cables may not be suitable and other plastics are preferred, notably polychloroprene (p.c.p.), which is now specified for high-class farm installations.

P.C.P. is a rubber-like material with an insulation which is proof against the highly-corrosive farm products and also against weathering.

Strong ozone, which is formed by the direct action of the sun's ultraviolet rays, tends to crack p.v.c., whilst p.c.p. cables are unaffected by this. The sheathing should preferably be coloured black for outdoor use.

Twin-sheathed cables usually have red and black cores. Are they permissible at switch positions?

The use of black- or blue-coloured wires at single-pole switch positions is generally an infringement of the IEE Wiring Regulations, yet it is a most common feature. In order to comply with the Regulations either a pair of red wires must be used which, contrary to common belief, is obtainable, or the black wire should be ringed with red sticky tape as fitted to m.i. cables.

What special precautions must be taken in explosive environments?

All apparatus for these applications must be *flameproof*, and there are two types:

(1) Mining gear used solely with armoured cable and special flexible cords.

(2) Industrial gear which may be employed with solid-drawn (seamless) steel conduit, m.i.c.s. cables, aluminium-sheathed or armoured cables.

All apparatus is normally of heavy cast-iron construction bearing the Government Buxton Certified symbol. A conduit barrier box must be positioned between flameproof and non-flameproof areas, the box interior containing a heavy insulating barrier with terminals to prevent the explosive gases from travelling right through the conduit system.

Carrying out flameproof work requires specialized treatment. The explosive gases have been divided into groups. Under Group 1 is firedamp, containing the combustible gas methane (CH_4) which is present in coal gas and causes explosions in coal mines. Fine coal dust can also explode when mixed with air.

Industrial explosive gases, vapours and inflammable liquids come within Groups 2 and 3. The explosive dusts may be either metallic (magnesium, aluminium, silicon, zinc and ferro-manganese) or of organic origin and where doubt exists advice should be sought from Government Factory Inspectors.

As we have seen, the standard system for house wiring is the use of twin and earth sheathed cable with its 3-plate connections. Are there any future possible improvements?

3-plate wiring which entails looping from point to point is certainly wasteful in cable, and has the further disadvantage of bringing live conductors to the ceiling outlets in addition to a bunch of cables. A simple improvement which suffers from none of these shortcomings would be the use of bakelite joint-boxes. Unfortunately they are positioned under floors. With the present-day fitted-carpet floor covering, the IEE Regulations would condemn these box connections as being in an inaccessible position.

Fig. 52 House wiring junction box (*Electrical Advisory Service*)

Recently much thought and work has been given to finding an improvement to the random positioning of joint boxes by a purpose-made large junction box with revolutionary printed circuit connections rated at 10 A. By placing the box (Fig. 52), or boxes, in the 'centre of gravity' (normally close to the concentration of room switches) of the installation, the system would appear to have merits of both simplicity and economy. However only by experience in the field can it positively be stated whether such a new method has any gain in cost and time in comparison to existing techniques.

Does traywork come under the category of a wiring system?

Traywork, although by itself is not a wiring system, follows many of the conduit and metal trunking techniques. The basic form consists of lengths of slotted metal with upturned sides (Fig. 53). Standard types are manufactured from a mild sheet steel alloy 3.5 mm ($\frac{1}{8}$ in.) thick and supplied in lengths of 2.4 m (8 ft), with a hot-dipped galvanised finish, to produce a continuous chassis for supporting and fixing cables. Slots in the tray make for easy fixing or re-arrangement of wiring.

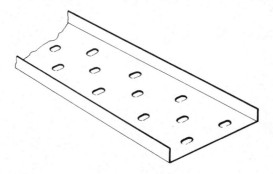

Fig. 53 Slotted cable tray

Added fitments include internal and external bends, tees and cross-tee pieces. Other finishes are self-colour, painted red oxide, yellow chromate or plastic-coated. The heavy-duty pattern when fitted mid-span — being a characteristic feature of cable traywork — takes loads up to 150 kg per metre run.

An interesting feature, which greatly extends the usefulness of this method, is an accompanying special bending machine. It has a lever somewhat similar to that required for conduit bending machines and can form the straight tray into angle, double or saddle sets.

If a ceiling rose is positioned between two ceiling wood joists, what is the method of fixing?

With the conduit system of wiring, little difficulty should be encountered as M4 metal screws will support the accessory by turning directly into the tapped lugs of the conduit box.

Toggle bolts are often employed for wiring consisting of twin and earth p.v.c. cables. This method, although quick and simple, is not recommended since any appreciable weight of the luminaire is bound to strain and perhaps damage the ceiling board.

By nailing a solid strip of wood approximately 200 mm long to the side of each joist and then fixing a flat board — often called a "shoulder" — to the strips, the boards then acts as a solid support for the ceiling rose. This method avoids any direct pull on the ceiling.

Chapter 8
The Ring Circuit

What exactly is the ring circuit?

This is a final circuit and therefore it is incorrect to give it the name — as is often done — of a ring main circuit. A clear definition may be found in IEE Regulation A35, 'Each circuit conductor of a ring final sub-circuit shall be run in the form of a ring, commencing from a way in the distribution board (or its equivalent), looping into the terminals of socket outlets and joint boxes (if any) connected in the ring and returning to the same way of the distribution board'. Socket outlets, which may be switched or unswitched, are of the 13-A type. The ring-main principle, whilst largely used in heavy industrial installations, and in fact employed in the National Grid, certainly proved an innovation when applied to the domestic field.

Why the *13*-A size? Here it has proved a lucky number: 13 A × 240 V (standard supply voltage) = 3 120 watts, which is sufficient for most portable appliances.

The plug is made to BS 1363 which specifies strict requirements for a good-quality product with a definite side-entry to discourage any pull on the flexible-cord lead when withdrawing the plug.

How has the 'all-purpose' nature of the 13-A plug been achieved?

The former 2, 5 or 15-A size plug with a choice of two or three pins has been replaced by the newer plug. The secret lies in the plug fuse: BS 1363 originally specified a choice of 2, 5, 10 and 13-A fuses so that the fundamental safety principle of the fuse matching the load could be achieved. Unfortunately, users could not be trained to fit the correct size of fuse, so that plugs almost invariably contained the 13-A fuse irrespective of the load. As a compromise the latest Standard Amendment specifies only the 3- and 13-A fuse, in the hope that the smaller size will be fitted to control loads up to 450 watts.

In addition to standardization are there any further advantages?

There is a clear saving in wiring costs due to diversity, the latter being a term used to express the fact that maximum demand is usually less (and can never be more) than the total connected load. In electrical installations cable sizes should be kept down to the minimum consistent with safety and are therefore based on maximum electrical demand of all appliances rather than total connected power as given in watts. Due to diversity, one 30-A ring circuit is permitted for all socket outlets covering an area of 100 m^2 (formerly 1 000 ft^2). It is interesting to note that the maximum load demand for 30 A is equal to 30 A x 240 V, i.e. 7·2 kW.

The cable size required for wiring the circuit is 1·5 mm^2 m.i. cable, 2·5 mm^2 p.v.c.-sheathed, and 4 mm^2 for copperclad. Care is required to ensure that the ends of these single-stranded cables are securely held in the socket-outlet terminals. The length of cable ends should be formed so that there is no undue pressure on the terminals as the socket outlet is screwed into position.

The radial system of wiring requires each 15-A point to be wired back to a separate way on the fuseboard. Thus, in addition to the need for added cable runs and all that this involves, extra fuse gear and switchgear is necessary. The ring method has also encouraged installation engineers and manufacturers to initiate ingenious new developments — a process which is still continuing — allowing for increased scope and making for more economical and improved installations.

Furthermore, the ring circuit has helped towards electrical safety in the home. The expensive old system often led to makeshift arrangements which could lead to electric shocks and the outbreak of fires. The 30-A circuit, by allowing the liberal use of socket outlets, has at one blow been able to prevent the widespread misuse of multiple adaptors, which often resulted in overheating and incorrect excess-current protection of appliances. Where 13-A adaptors are unavoidable, fused types are available in a variety of forms.

How many outlets should be provided in dwellings?

Speculative builders in the past have often given as little as one per room, resulting in a total of about seven per installation; American journals reveal the other extreme, up to 85. Nevertheless, this figure is indicative of future trends and on this basis 30—40 socket outlets are not excessive for a medium semi-detached three-bedroomed house when allowance is made for any additional labour-saving devices which may be expected.

Installation engineers regard flexible cords as weak links — in contrast to the main-circuit wiring. The liberal supply of socket outlets minimizes undue lengths of flexible cords and thereby becomes another major factor making for safety.

Where should the outlets be sited?

It is a mistake to fit the points too close to the floor where they would be vulnerable to kicks, and where the cord outlets from the plugs make them subject to sharp bends. The latter is liable to produce a strain on the cord grip and a weakening of the flexible cord strands at this point. For normal use a minimum height of 150 mm to the bottom of the fitment is recommended except for kitchens, where they should be raised to 1·2 to 1·35 m. For best planning, careful consideration must be given to the routing of the runs so that the excessive current-consuming appliances are not all connected to one side of the ring. The most favourable conditions for the loading of the circuit takes place when the electric fires are plugged in so as to balance the circuit on both 'legs' of the ring. An unbalanced condition is sometimes the reason for the apparently mysterious heating of the first leg from the 30-A fuse, a situation which can arise if this cable is routed up the staircase and is therefore much longer than the feed from the other side.

How should the cable runs be laid out?

In order to economize with cable and to avoid wasteful doubling back, in contrast to radial circuits, much more care is required in planning the ring-circuit layout. A layout of points is first drawn up and it is a good idea to give a number to each of them on an architect's or, if not available, an improvised plan of the house.

Even where the floor area is less than 30 m^2 it is now usual to fit at least two ring circuits. However, the obvious method of running one ring circuit for the ground floor and one for the first floor is not always practical because if the ground floor is of the solid type it would entail expensive cable protection. A first floor with wood joists should satisfactorily accommodate the necessary 2·5 mm^2 ring-circuit cabling except for the vertical spur runs.

What are spur points?

These are, in effect, points which are tee'ed off the main ring circuit (Fig. 54) either by a loop from one of the sockets or by means of a joint box, usually resulting in a saving of time and labour. The cable feeding spur points must be of the same rating as the conductor serving the ring.

Each spur may consist of one or two 13-A points (or one twin

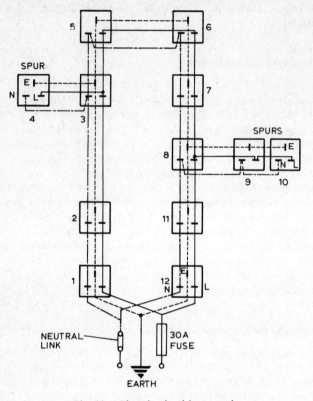

Fig. 54   Ring circuit with spur points

socket outlet) but only one stationary appliance is permitted to be fed from the non-fused spur.

It is important to note that the total number of fused spur outlets must not exceed those on the ring itself.

What is a line diagram?

Referring to p.v.c.-sheathed cable each line in the diagram represents a twin and earth cable, advantage of which is taken in planning out the routes. Fig. 55 should now be seen in conjunction with the points as set out in Fig. 54, the numbers in the small circles corresponding to the points marked on the plan. In this way, by following the runs it is easy to see that each ring circuit consists of a live phase wire starting from the 30-A fuse, feeding each socket in turn, branching off for spurs where necessary, and returning from the last socket to the same terminal of the fuse, the neutral and earth conductors following in the same manner.

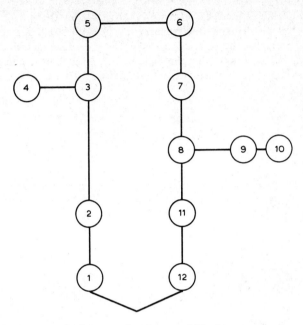

Fig. 55   Line diagram of Fig. 54

Where the ground floor is solid the house is divided *vertically* for electrical purposes, each of these halves occupying one of the ring circuits. To minimize weakening of the building most crossing of the joints occurs near the walls, while notching should be avoided.

Can the 13-A plug be used on non-ring circuits?

Fig. 56 (*a, b* and *c*) depicts line diagrams showing how the points can be wired to follow a radial circuit in at least three different ways and still comply with the IEE Regulations.

In the first example (Fig. 56 (*a*)) there is no economy; this circuit simply allows the convenience of local fusing, but the arrangement suffers from lack of 'discrimination', i.e. a large reduction in fuse sizes as the wiring proceeds from intake to the final sub-circuit. This is a major factor in circuit planning and is often overlooked. The 13-A fuse in the plug is little different in current size from the 15-A circuit fuse and therefore, in the event of a short-circuit, the 15 A may easily blow before the 13 A. An improvement is to use 20-A cable and replace the 15-A fuse by one of 20-A rating.

The circuit in Fig. 56 (*b*) a distinct improvement since both socket outlets may be wired with 2·5-mm^2 cable. This is made possible because, by taking advantage of 'diversity', it is unlikely that each of the socket outlets will be used simultaneously at their full rated capacity. It can also be seen that the 20-A fuse gives much better discrimination.

By using heavier-gauge cable throughout (Fig. 56(c)), the circuit may be extended up to six points when controlled by a 30-A fuse. This may

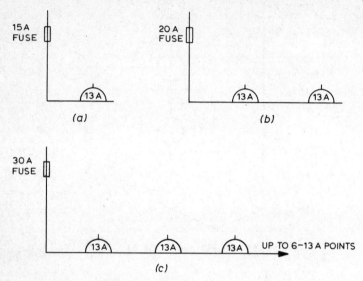

Fig. 56 Use of 13-A outlets in radial (looped) circuits

effect considerable economy in a long narrow building where the sockets are a long way apart, and the arrangement can be fitted into a grid system of wiring (which spans the interior of the building by means of a steel-stranded catenary wire).

Are additions simple to make?

The simplest method is to replace any socket outlets by the 2-gang type which does not entail any extra wiring. 'Back-to-back', i.e. points on both sides of a partition, are also possible by means of a long backless square box which allows plenty of room to accommodate the necessary cables. Care must be taken, however, against overloading the circuit; the regulations are perfectly clear in defining the twin unit as two outlets and not one point.

Where runs allow, 'teeing-off' as a means of feeding spur points are permitted and the larger 30-A bakelite joint box, conforming to BS 816, is especially manufactured for this purpose. One type has a selective entry feature which closes unused entry slots in the base.

A great deal of ingenuity by electrical-accessory manufacturers has gone into designing conversion units. In one type the twin unit is fixed by two long 3·5-M screws registering into the same 60·3 mm centred, single-gang, tapped screw-holes of the flush box.

Plastic hollow skirtings are also available, allowing maximum flexibility and re-arrangement, at any time, of the home, office and work furniture. A further point to note is that if the appliance is non-luminous, such as a storage heater, pilot lamps are an advantage.

How is the current in each section of the ring calculated?

It is easy to calculate the current feeding any particular radial socket outlet but for the ring circuit it is not so straightforward. While such calculations may not be so essential (since protection against overloading is provided by the 30-A fuse), there are occasions when one wishes to analyse why the cables on one side of the ring are getting hot.

For calculations the use of Kirchhoff's Laws, which may be seen as an extension of Ohm's Law, is required:

(1) When currents meet at a junction, the algebraical sum of the currents is zero.

(2) In any closed circuit, the sum of the p.d.s. is equal to the sum of the e.m.f.s.

How is testing for insulation faults carried out?

Due to the inherent inter-connection, testing for earth faults can be a lengthy process. Obviously the faulty section should be isolated but there is no short-cut to achieving this aim, and a certain amount of skill is required in searching for a cable damaged by a nail.

Probably the best method is to split the ring circuit and test both halves and this procedure should be repeated until the faulty portion is located.

By now it must be apparent to readers that the ring circuit can have anything but a circular shape. The importance of retaining a line diagram of the cable runs will be appreciated as a great time-saver. Being in the form of a chart it is also valuable when extensions are required, in order to ensure that the correct ratio of ring to spur outlets is maintained to avoid overloading the cables.

Has the 13-A plug come to stay?

In spite of its many advantages, the 13-A accessory is a native product. To achieve international interchangeability, moves are afoot to produce a standard plug with a size of 16 A, so that at 240 V it would still have a rating of more than 3 kW.

Another unusual feature is that the fuse would probably be contained in the socket portion.

Chapter 9
Installation Accessories

What is meant by 'accessory'?

Probably the clearest definition is given in the IEE Wiring Regulations, 'Any device, other than a light fitting, associated with the wiring and current-using appliance of an installation, e.g. a switch, a fuse, a plug, a socket outlet, a lampholder or a ceiling rose'. In other words, parts which are necessary to complete an installation and usually common to all wiring systems.

What type of box houses the working parts of flush switches?

Fig. 15 (p. 24) depicts the standard square box conforming to BS 1299 (now BS 3676 which contains requirements formerly in BS 1299) complete with earth terminal. This box has perhaps helped to make the most progress towards standardization of accessories as it will accept the 2-gang switch unit (in some makes the 3-gang switch unit), the 13-A switched-door or unswitched socket outlet, spur boxes and a host of accessories which have a distance between fixing-hole centres of 60·3 mm. The holes are tapped 3·5 M and the fixing screws are normally of the nickel-plated raised-head pattern.

The sides of the boxes have 'knockouts', circular plates which may be knocked out to accept mineral-insulated glands or a variety of conduit bush sizes, 16, 20 or 25 mm. Rubber grommets make for a definite economy where entry to the boxes is by p.v.c.-sheathed cables.

With the same front dimensions, deep, medium or shallow boxes are available. Knockouts may be replaced by cleats for insulated and unscrewed conduits and the shallow metal boxes are often supplied with elongated p.v.c.-bushed holes.

How are boxes fixed to hollow walls or dry partitions?

Hollow walls are found in older properties; the new dry, thin partition wall completely eliminates any need for plastering and is in line with the

Installation Accessories   81

general trend in building to cut costs. Unfortunately, fixing flush boxes under these circumstances presents problems as there is no thickness in the wall to accommodate any fixing screws. To overcome such difficulties a box with special lugs has been designed (Fig. 57) which is also suitable for hollow walls with plasterboard and other facings.

Fig. 57   Tenby box for hollow walls   Fig. 58   Finger-touch dimmer switch
(*Tenby*)   (*MK Electric*)

Are boxes used for surface switches?

The old-fashioned round switches for mounting on wood blocks, which must be of hardwood such as beech, oak (English), teak or mahogany, are still listed in manufacturers' catalogues. It is, however, more usual to give installations a modern look by fixing flush switches on to plastic pattress boxes. For plastic-conduit work the box may have a metal reinforcing piece.

Flush switches may also fit into specially-shaped metal boxes for surface finish. In factories the switches will perhaps be provided with metal plates, although the switch knobs will probably be of an insulated non-combustible material.

What kinds of switches are available?

The average householder may perhaps be forgiven for thinking only in terms of the 1-way or 2-way switch. The following, which is by no means comprehensive, is given as an indication of the possible range available:

    1-way single pole   2-way
    1-way double pole   intermediate

twin knob 1-way and 2-way
series—parallel
push button

key operated
thyristor dimmer switch
(Fig. 58)

Switches are obtainable with dolly knobs although the rocker mechanism appears to be increasing in popularity. The switches listed above all fit into the standard switch box; in many cases 5- or 15-A ones may be selected. Certain other special switches for particular circuits, such as the double-pole series—parallel type, can be obtained to order.

Multi-gang switches up to 24 ways are also available, these being mounted on a grid to make up the requisite numbers. These units lend themselves to imaginative siting of switches, e.g. a bank at the fireside allows for control without the need to move across the room.

In appropriate circumstances the architrave oblong-plated switch forms an attractive alternative to the relatively large, square, standard plate. Here vertical mounting is necessary for the 2-gang type.

Colouring for plastic switch plates has settled down to white or off-white, with occasional use of brown. It is still possible to obtain some choice of stamped metal plates (which must be earthed) — bronze, steel grey, polished chrome or matt chrome.

For rooms equipped with period furniture, heavy-lacquered, solid brass switch plates are eminently suitable, fixing being by the old-fashioned ring which threads on to the brass switch-knob.

How are accessories fitted to trunking?

Since the trunking is, in effect, a long hollow box, simple adaptable mounting frames may be all that is necessary for fixing square switch plates or socket outlets, etc. although some engineers may prefer the additional cable stowage space and other advantages offered by the surface-mounted box.

Why is the slow-break switch preferred to the quick-make-and-break type?

With an alternating current, the sparking at break and consequent burning of the contact is less if the break is slow and short rather than sharp and long. A word of warning has to be given here as this action may not always be true of highly-inductive circuits, although it must be stated that at least one manufacturer strongly recommends their use for fluorescent lighting. Switches not designed for inductive circuits are required to be run at half their current rating.

The development of the slow-break switch for a.c. use has several other beneficial effects: the noise associated with the quick-make-and-break switches has been practically eliminated. A considerable simplification of design has resulted. The complicated toggle action has disappeared, although to some extent small springs have been retained, mainly for securing good contact and, by being always in compression, ensuring long life. Switches of this type are rated up to 80 A at 250 V a.c., and are even in common use on 415-V 3-phase supplies for low-powered motors.

What are other advantages of the micro-gap switch?

Designers have made it possible to evolve even simpler switch actions with the slow-break micro-gap switches; in one type the only moving part is a beryllium—copper grid, forming its own contact bridge. Modern techniques are incorporated in making the actual contacts of silver, as this is the best of all conductors and serves to reduce heat which would otherwise encourage ionization or arc formation. Further, a sliding action is imparted to ensure automative cleaning of the contact faces during the switch operation. With these newer types there are none of the corrodible or highly-stressed components found in older types of switches, neither is there any *contact bounce* arising from heavy switch action. Another advantage is enclosure of the movement which thereby protects it from dust and atmospheric conditions.

When micro-gap switches, which were originally designed for a normal load of 5 A, were tested at a current of 10 A, the useful life was shown to be upwards of 100 000 operations (at 50 operations a week this would indicate a minimum working life of about 40 years). This compares favourably with the British Standard requirement which specifies 15 000 operations at the normal current and voltage.

What is the difference in use between the single-pole and double-pole switch?

For general electric-light control and 13-A switched socket outlets, single pole switches are generally employed. *The switch must be connected in the live side* of the supply.

Crossed polarity or the inadvertent interchange of the L and N conductors leading to a switched neutral may be lethal. The danger of such a situation occurs when an exposed element is being cleaned or otherwise touched by hand, while the other hand is in contact with the metal casing. Double-pole switching could eliminate this hazard, although some people advocate complete withdrawal of the plug, where possible, as a certain safe form of action.

The regulations specifically state that, where an appliance is fitted with heating elements that can be touched, or into which more than one phase of the supply is introduced (e.g. 3-phase apparatus), the switch must be of the linked type and designed so as to break all circuit conductors, simultaneously. This type of control is also necessary for infrared reflector fires with silica-sheathed elements, owing to the fragile nature of the sheathing. Furthermore, the switch must be in a reasonably accessible position and a circuit-breaker may be used as a switch.

For temporary installations where the circuits are connected to 2-wire supplies, use should be made of double-pole switches and/or linked circuit-breakers. This is essential for situations where neither pole of the supply is earthed, and is desirable in any circumstance where the ability to maintain correct polarity of the supply is at all in doubt.

Why are shaver-socket outlets not earthed?

Electric shavers form an exception in the special case where used in bathrooms or adjacent to washbasins. The shaver-switch socket unit must conform to BS 3052 and incorporate a 1:1 isolating transformer ensuring that neither lead connected to the actual shaver head is at earth potential, since here the neutral/earth link no longer exists. Clearly, even if any conductor in the shaving device became detached and were to touch the metal part with the user in contact with this faulty part and earthed metal pipe he would not form part of an electric circuit. Thus, under these circumstances an electric shock could not be received.

What type of switch control is necessary for the electrode boiler?

The requirements are for a double-pole linked switch or its equivalent, which is separate from and within easy reach of the heater or boiler (the wiring from the boiler must be directly connected to the switch without the use of a plug or socket outlet).

The electrode boiler requires special care in installation as it may form an electrical hazard, the reason being that, in contrast to the immersion heater, uninsulated electrodes are in direct contact with the water. The resistance of the water becomes the 'element' as a.c. is passed through it.

Where should the positions of switches be in relation to the loads they control?

In general the controlling switch should be in a readily accessible position and as near the load as possible. Every switch or circuit-breaker, the purpose of which is not obvious, needs to be *labelled* in order to indicate the apparatus it controls.

Again bathrooms call for care and switches must be sited so as to be inaccessible to a person using the bath. For this reason light switches are often placed not at the bathroom door but outside it. It is, however, more common to fit ceiling cord switches inside the bathroom itself.

Agricultural and horticultural installations also require special precautions: every switch or other means of control or adjustment, not forming an integral part of other apparatus, must be so situated as to be out of reach of any person in contact with wash troughs, sterilizing equipment and the like.

What are the difficulties in connecting metric cables to accessories and how can they be overcome?

As has already been mentioned, certain difficulties have arisen regarding the use of single-stranded 2·5-mm^2 cable with 13-A socket outlets, and to simplify these the cable has now been made available in a 7-stranded form.

The stowing of spare cable has always been the hallmark of the careful worker, and is particularly necessary for the metric cable. The short tails of solid-core cables are too rigid to enable the accessory to be pushed back into the box without undue strain on the accessory and connections and it is recommended that the cable be cut so as to allow a length of 127 to 152 mm to loop within the box.

Outer sheathing of p.v.c.-sheathed cable should be stripped back, inside the box, near to the knockout or cable entry; it is also advisable to arrange the cores to conform with the terminal positions of the socket outlet or similar accessory. When fitting stranded cables it is standard practice to twist together the bare ends of the conductors prior to securing them in the terminal; this method may not be practicable with solid conductors where it becomes only necessary to insert the bare ends fully into the terminals. Some switch designers claim that as an alternative to the clamp-type terminals large-diameter screws will grip and hold the ends securely in good permanent contact. The accessory can now be pushed gently back into the box.

A point to note is that the distance from the back projection of the moulded accessory to the back of the box should be at least 14·3 mm to allow for the stowage of slack cable.

What is maximum voltage for ceiling roses?

Normally not above low voltage, the maximum being 250 V. This itself is an 'r.m.s.' value so that the accessory may be subjected to a pressure of $\sqrt{3} \times$ 250 V, i.e. 433 V, even without allowing for the permissible statutory variation of plus or minus 6 per cent of the supply voltage.

What types of ceiling roses are available?

There is a basic change between the older types and present-day fitments. For very many years the two-plate 'china' (porcelain) ceiling roses did not undergo any change in design and many are found to be still in service on older installations, and even the replacement by bakelite made for little change at first. The 1955 13th Edition of the IEE Regulations did not permit the roses to be fixed directly to ceilings without any proper back enclosures.

The Regulation (207H) which brought this change is worth quoting in full as it still forms a basic requirement of modern installation thinking:

> Cores of sheathed cables from which the sheath has been removed and non-sheathed cables at the termination of conduit or duct or trunking shall be enclosed. The enclosure shall be of incombustible material and may be a box complying with BS 816 or other appropriate British Standard or an accessory or lighting fitting. Alternatively, it may be formed by part of an accessory or lighting fitting and the building structure. In a damp situation the enclosure shall be damp and dust proof and corrosion resistant.

Note: The requirements apply particularly where cables terminate at, or are looped into, an accessory or light fitting.

Nowadays the vogue is towards combining the pattress and ceiling rose into one unit; a contemporary styling with the 'Halo' effect also minimizes the need to fill in any gaps round the ceiling rose. Added points to note are that each of the earth-continuity conductors from p.v.c.-sheathed cables should have an insulated sleeving coloured yellow/green; and a further safety feature is introduced by the insulated shrouding of the live-line loop terminal. It is also interesting to note the various means adopted for the cord grips which take the strain off the flexible-cord connection.

In yet another type the terminal blocks are angled with the aim of easing the wiring process: there is a separate terminal for each cable thus eliminating the need to twist solid-core metric conductors. Here *all* terminals are enclosed by a transparent plastic block (Fig. 59). A further desirable feature is provided by a choice of various types of strain clamps in addition to the normal cord grip.

An accessory which goes under the title of 'ceiling master' avoids the need to disconnect the wiring to light fittings when the latter may have to be taken down for the purposes of cleaning or maintenance. The upper part, which is fixed permanently to the ceiling, contains contacts connected to the mains, while the lower part, which carries the light fitting, slides into this upper section.

What further changes are taking place in wiring design?

Changes are bound to occur in line with developments in electrical installations, many of which in turn are brought about by the need to comply with alterations in constructional techniques.

An example is what is known as industrialized building as used for the construction of tower blocks of flats and, since the work is of a repetitive nature, the wiring lends itself to the use of certain of the mass-production factory methods. The aim of the wiring system in these circumstances is towards a simple harness as employed in the automobile industry. Tabbed

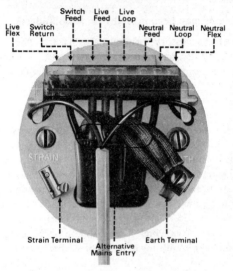

Fig. 59 Rock ceiling rose

cables with made-up ends are being used to save the time and labour on site. One further simplification which is advocated is to run the domestic lighting supply off the ring circuit, for which ceiling-rose outlets have been devised with a type of 5-pin socket and plug which houses a 3-A fuse and is interposed between the 2·5-mm ring-circuit cables and the 1·5-mm lighting-switch feed cables.

What precautions are necessary when fitting lampholders in damp situations?

In damp positions or where lampholders can readily be touched by a person in contact with, or standing on, earthed metal, then the lampholder itself — if of metal — must be definitely earthed. Alternatively, the all-insulated type which conforms to the appropriate standard and requires a protective shield or skirt may be fitted. For bathrooms totally enclosed fittings are recommended for obvious safety reasons. Irrespective of the situation, lampholders may not normally be used on a higher voltage than 250 V.

How should connection be made to the Edison-type screw lampholder?

While replacing a lamp into a screw-in type of lampholder, contact by the fingers may easily be made with the lamp threads. For this reason the outer or threaded portion of the holder must be connected to the neutral conductor; the same applies to centre-contact bayonet lampholders.

What other problems may be expected with lampholders?

Excessive heat at the accessory forms a major hazard. The filament or incandescent lamp by its very nature is bound to run very hot so that, although it is by no means a new problem, we are now faced with an intensification of heat. Tungsten was eventually chosen as a lamp-filament material because of its extremely high melting point, some $3\,380°C$ — it is operated in some of today's lamps at near $3\,000°C$.

Although the increased efficiency of present-day electric lamps is clearly a result of higher filament temperatures through the use of coiled-coil lamps, this is by no means the only factor. The temperature rise is augmented by the actual decrease in the lamp dimensions.

At these elevated temperatures the ordinary cheap, moulded lampholder can be rendered useless, since the temperature even at the shade ring can be as much as $135°C$ and this value is clearly the limit for the cheap brown material, after which it shrinks and becomes brittle. In the early days no cheap moulding capable of withstanding such high temperatures was found.

Brass lampholders, which were almost all replaced en masse by the insulated types in the 1930s, withstand these high temperatures and there was no problem with jambed shade rings; for these reasons they are sometimes specified by consultants. However, 3-core flexible cord would be necessary to earth the holders even though the shades may be of plastic.

The intense heat of the coiled-coil filament has brought the temperatures at holder terminals to $165°C$, and so the phosphor-bronze springs can no longer be used for the plungers. Ordinary steel, which may be used up to a maximum temperature of $170°C$, has replaced the earlier used metal; alternatively, the springs may be made of special steels which will withstand a temperature of more than $200°C$.

What type of lampholders should be used for pendants?

An IEE Regulation stipulates that the insulated, moulded bayonet-capped (BC) lampholder must comply with BS 52A. Where the accessory is likely to reach a working temperature of $135°C$, it must be of the heat-resisting type and made to BS 52H.

A further note, 'For compliance with Regulation C24, the use of heat-resisting lampholders is likely to be necessary in all cases where lamps are installed in the cap-up position, or where they are contained in an enclosure fitting or inadequately ventilated shade'. From this it follows that to comply with the IEE Regulations, holders to BS 52H are necessary for very many light points.

An obvious answer would be the avoidance of spring plungers altogether and the use of the Edison-screw-type (ES) lampholder for the small wattage lamps as normally fitted in America and the Continent — oddly enough on a recent visit the increasing use of BC types was noticeable — but this would entail rejecting long-established customs, except for the 150- and 200-W sizes where *either* BC or ES caps are obtainable.

Above the 200-W size Edison screwholders are standard. For the large-wattage filament lamps, 500–1 500 W, Benjamins developed, some years ago, their Saaflux system, the principal features including the top supplementary reflector which can be seen below the lampholder close to the neck of the lamp, preventing convection currents carrying heat from the lamp filament up to the lamp cap and lead-in. The lampholder is of simple construction and consists of an open shell so that the heat can escape from it. The ribbed rods supporting the lampholder hold it rigidly in place at the correct focal point and are also so constructed so as to radiate heat and to prevent too much of it from reaching the upper part.

Where the top cup joins the reflector it is sealed to prevent heat rising to the lead-in cables and also to stop dust from entering the top of the fitting. The top cup fits into a flange to which it can easily be detached and as easily removed, and is locked by means of a knurled nut; this enables the entire fitting to be removed for completely cleaning the lamp.

The top flange also carries an insulated wiring block to which the cables are connected. When wiring, only this flange is installed, the fitting being inserted later, and this is all that the electrician needs to support when making the connections, in addition to the connecting block which weighs less than 0·085 kg. The electrical connection is made to the terminal block by two blades which make a firm fit with phosphor-bronze contacts, and these can only be inserted one way round, thus ensuring correct polarity.

Reverting back to the common BC plastic lampholders, of which there is a choice of eye-catching modern styling conforming to BS 52H, certain makes ensure the locking of the top half which covers the flexible cord connections. Without this arrangement there is always the danger that, when taking off a lampshade, the top half can be inadvertently unscrewed, thus allowing the live terminals to be touched. A further point is that coarse threading reduces the possibility of thread binding when unscrewing shade rings.

It must also be seen that all filament lamps must either be positioned or guarded so as to prevent ignition of any flammable materials. Shades or guards must also be designed so as to withstand heat from lamps.

Where may neon indicator lamps be fitted?

Immersion and storage heaters are two items of electrical equipment which do not give immediate indication, by some form of pilot light, of being energized. For this purpose the neon indicator, with its long life and low current consumption, fulfils a valuable function.

One form may be housed in a 13-A switch socket (Fig. 60) and its small physical dimensions permit the lamp to be fitted in a 13-A plug. For storage heaters a common practice is to fix the indicator in the flex-outlet unit.

Fig. 60   Neon lamp indicator
(*MK Electric Ltd*)

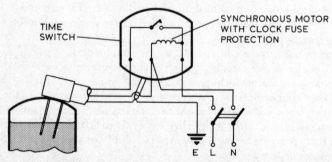

Fig. 61   Time-switch control for immersion heater

What are time switches?

They are, in essence, motorized switches where the synchronous motor is geared and calibrated to drive an electric clock. Levers, which form part of the mechanism, are designed to switch circuits on and off at predetermined times, an obvious application being for shop-window lighting. In domestic appliances it may be wired to an immersion heater (Fig. 61) which is connected to an off-peak or white-meter supply. By this means an early morning tankful of hot water can be obtained at a cheaper rate.

When connecting p.v.c. sheathed flexible cords to 13 A plugs, what special precautions should be taken?

(1) Special care is necessary to avoid cutting into the core insulation when stripping the outer sheath. For cutting into the sheath, the knife is held at a very slight angle. Craftsmen produce a neat finish by employing

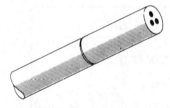

Fig. 62 Circling p.v.c. sheath to produce a neat finish

the knife to circle a very small amount into the sheath (Fig. 62) so as not to damage any core insulation. This indented circle is made at the place where stripping has commenced; finally the surplus sheathing is pulled away.

(2) Core conductors must be cut to correct lengths. Exact dimensions will vary with different makes of plugs. Allowance must be made for the sheath to enter well into the mouth of the plug — but not too far!

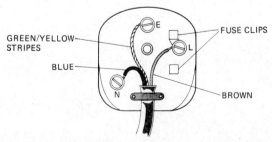

Fig. 63 13 A plug connections

92   *Electrical Installation*

(3)   Colour coding must be observed. Brown for live (L), blue for neutral (N), and green/yellow stripe for earth (E) as shown in Fig. 63.

(4)   Where a copper stripped end is to be wrapped round a terminal screw, the strands should be tightly twisted together, normally between thumb and forefinger, and then formed round the blade of a thin screwdriver. Finally the strands are tightened under the washer. With the pillar type of terminal, care must be taken to allow for doubling over the stripped conductor. For either terminal, no exposed copper conductor should be shown.

(5)   Before fixing back the plug cover, ensure that the flexible cord sheath is properly held by the plug cord grips.

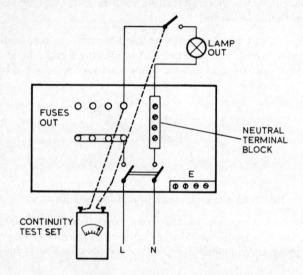

Fig. 64   Test for correct polarity

Chapter 10
Testing

What is the purpose of testing?

The main purpose of testing is to detect faults before dangerous situations arise. As a minimum, installations must be protected against the following possible hazards: earth leakage, electric shock, excess current, moisture and corrosion. This will require tests both by instruments and by physical inspection.

When should the tests be carried out?

(1) Before a new installation is put into service.
(2) When additions (especially major) are made to an installation.
(3) Periodically, at intervals of not more than five years.

What is the recommended sequence of inspection and testing?

(1) Continuity of ring circuit.
(2) Continuity of earthing conductors.
(3) Earth electrode resistance.
(4) Insulation resistance.
(5) Insulation of site-built assemblies.
(6) Protection by electrical separation.
(7) Polarity.
(8) Earth loop impedance.
(9) Operation of current and voltage earth leakage circuit-breakers.

How should the polarity tests be carried out?

The main purpose is to make certain that the live conductors and not the neutrals are connected to the switches; Fig. 64 shows a simple method. For this test the supply must be disconnected or switched off and all switches must be in the off position; at the same time all fuses taken out (miniature circuit-breakers in the off position) and neutral links kept in place.

The polarity test must be applied to all plugs and sockets, in addition to centre-contact bayonet- and screw-type lampholders to check that the outer contact, in each case, is connected to the neutral conductor.

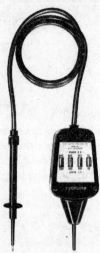

Fig. 65 Test prods and supply indicator (*Martindale*)

The common practice of using a neon tester for polarity testing with the supply on is to be deprecated as it can lead to misleading results and may be dangerous should the device become faulty. It is preferable to use a purpose-made test lamp with suitable safeguards, such as a guard over a rough service lamp and h.b.c. fuses in the test-lead prods (Fig. 65).

What are the basic earth-test requirements?

To test that, in the event of a live conductor inadvertently touching earthed metal, the fuse will blow or the excess-current circuit-breaker cut out.

More specifically, the test will reveal that the protective devices will operate if the value of the earth-fault current exceeds:

(1) 3 times the current of the semi-enclosed fuse or any cartridge fuse having a fusing factor of more than 1·5; or

(2) 2·4 times the rating of the cartridge fuse having a fusing factor of 1·5 or less; or

(3) 1·5 times the tripping current of the overload circuit-breaker.

The fusing factor is given by the ratio:

$$\frac{\text{Minimum fusing current}}{\text{Current rating of fuse}}$$

How is the main earth test carried out?

For most installations the major requirement is the loop impedance test; there is, in fact, a choice of two tests (Fig. 66). Before either are carried out appropriate cross-bonding of the consumer's main earth terminal, by a copper conductor of minimum size 6 mm^2, to the metalwork of any gas or water services on the consumer's premises must be carried out.

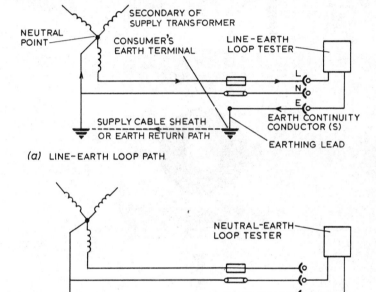

Fig. 66   Earth-loop impedance tests

In one type of line-earth loop impedance tester (Fig. 67) a current of about 20 A is allowed to pass through the entire path for about 30 – 50 milliseconds and the voltage drop across a resistor in the test instrument gives the value of impedance directly in ohms.

Neutral-earth testers are often known as injection-current types because, in order to operate, a current has to be injected in the neutral-earth loop preferably by a transformer fed from the mains. Alternatively the supply to the N/E tester may be by a d.c. source, the polarity of which is continually and rapidly reversing. This method of testing must not be employed when the system is earthed by protective multiple earthing (p.m.e.). Objections have also been made to its use as the values obtained by the instrument may be affected by currents flowing in the neutral supply conductor and by voltage changes in the neutral conductor itself.

Fig. 67   Line-earth loop impedance tester
(*Megger*)

What actual values of the loop impedance are necessary for the protective devices to operate?

In the case of a semi-enclosed fuse containing a 5-A fuse-wire, the maximum value of the earth-fault loop impedance on a 240-V circuit is equal to;

$$\frac{240 \text{ V}}{3 \times 15 \text{ A}} = 16 \text{ ohms}$$

With either current values, by similar calculations the following table may be obtained for the semi-enclosed fuse:

Current rating of fuse (amperes)	Maximum impedance (ohms)
10	8
15	5·3
20	4
30	2·7
45	1·8
60	1·35
100	0·8

If sufficiently low earth-loop impedance values cannot be obtained, what type of earth-leakage protection must be provided?

In all such cases, earth-leakage protection must be given by one or more *earth-leakage circuit-breakers*. There are two types, i.e. voltage-operated and current-operated earth-leakage circuit-breakers. The latter type is preferable because care must be taken to avoid shorting out the operating coil of the voltage e.l.c.b. while the current e.l.c.b. can be designed so as to minimize the possible effects of electric shock by limiting the values of any earth-leakage currents.

The earth-leakage circuit-breakers must themselves be inspected and tested before being placed in service and at subsequent regular intervals. In addition to checking the tripping test button or switch, a maximum of 45 V, obtained from a double-wound transformer, when applied across the neutral and earth terminals (or neutral and frame of the voltage-operated type) should permit the circuit-breaker to trip instantaneously. It is advisable that the short-time rating of the double-wound transformer be preferably less than 0·75 kVA.

If the maximum impedance value, as found by the loop tester, is too high then the various portions of the loop within the installation should be first checked. This will require individual tests for the earth-continuity conductors, the earthing lead and the earth electrode.

All these tests, some of which are specialized, are set out fully in Appendix 6 of the IEE Wiring Regulations.

What instrument is used for insulation testing?

The instrument is normally known as a Megger insulation tester. In a common form it comprises a hand-operated generator combined with a high-reading ohmmeter in one case, and when the handle is vigorously turned a test voltage of 500 V d.c. is generated. The ohmmeter has an extremely wide range from zero to infinity but the scale is not uniform, 1 megohm (MΩ or a million ohms) being at the centre of the scale.

There are a number of different types of insulation-resistance testers available, most of which may also permit continuity testing to be carried out; one type consisting of a push-button battery-operated megger may be obtained. Heavy-duty instruments (Fig. 68) are suitable for testing apparatus with a considerable degree of capacitance and a high degree of insulation.

Fig. 68   2·5-kV Major Megger tester
(*Megger*)

What insulation tests must be carried out?

Two tests are required, (i) between conductors, (ii) between all conductors and earth (Fig. 69), and must be carried out before the installation is placed in commission. *The minimum insulation value in each case must not be less than one megohm.*

For the test between poles, all lamps and all current-using apparatus — wherever possible — must be removed and switches placed in the ON position.

The earth test requires all poles to be connected together and again switches must be closed. For these tests large installations may be divided into groups of outlets, each containing not less than 50 outlets.

Clearly care must be required with earth-concentric wiring systems, as under these conditions the earth and neutral are not separate but form the same conductor.

For apparatus which cannot be disconnected during the above tests, separate tests may be made. The insulation resistance of each item of fixed apparatus must, when tested, not have a value of less than 0·5MΩ between poles or between all conductors and earth.

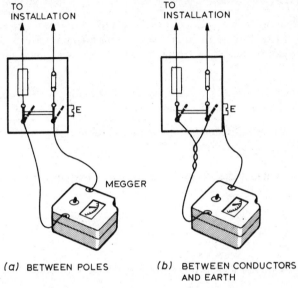

(a) BETWEEN POLES   (b) BETWEEN CONDUCTORS AND EARTH

Fig. 69   Insulation tests

What is the ring-circuit continuity test?

This is a test to verify that the line, neutral and earth-continuity conductors form a complete ring. It is usually carried out at the consumer's control unit and requires the two ends of each pole and earth to be disconnected in order to check the integrity of the ring.

Who should sign Completion and Inspection Certificates?

These certificates, which are required to comply with the IEE Wiring Regulations, should be signed by a competent person who should preferably be a professionally-qualified electrical engineer with installation experience.

By kind permission of the Institution of Electrical Engineers copies of the Completion and Inspection Certificates are given below:

COMPLETION CERTIFICATE
The insulation resistance of the fixed wiring installation is not less than 1 megohm (Regulations E.6–8).
The insulation resistance to earth of each item of apparatus tested separately is not less than 0·5 megohm (Regulations E.6 and E.9). Each item of apparatus tested separately is in good serviceable condition, except as stated below.
All flexible cords, switches, fuses, plugs and socket-outlets are in good serviceable condition, except as stated below.
There is no sign of overloading of conductors or accessories except as stated below.
Apparatus tested includes/does not include portable appliances.
Comments (if any) and departures from the Wiring Regulations:
Signed.. Date.........................
For and on behalf of: ....
....
Address: ....
....
....
....

INSPECTION CERTIFICATE

(as prescribed in the I.E.E. Regulations for the Electrical Equipment of Buildings)

Inspection Certificate to be given by the contractor or other person responsible for carrying out an inspection and test of an installation, or part of an installation, or by an authorized person acting on his behalf

I CERTIFY that the electrical installation at:

has been inspected and tested, in accordance with the requirements of Section E of the I.E.E. Wiring Regulations (14th Edition) and that the results are as indicated below.

I RECOMMEND that this installation be further inspected and tested after an interval of not more than...................... years.*

Items inspected or tested: †

Method of earthing: Cable sheath.
                              Additional overhead line conductor.
                              Protective multiple earthing (P.M.E.).
                              Buried strip/rod/plate.
                              Earth-leakage circuit-breaker, voltage operated.
                              Earth-leakage circuit-breaker, current operated.

The impedance of each earth-continuity conductor is satisfactory (Regulation E.3).
The total earth-loop impedance is satisfactory/unsatisfactory for ready operation of the largest-rated excess-current protective device relied upon for earth-leakage protection (Regulations E.3–4).
Earth-leakage protection is afforded by a current operated/voltage operated earth-leakage circuit-breaker, the operation of which is effective (Regulation E.5).
Polarity throughout the installation is correct (Regulation E.2).
All single-pole control devices are in live conductors only (Regulation E.2).

    *The space provided in the Certificate for inserting the recommended number of years intervening between inspections should be filled in with the figure 5 or such lesser figure as is considered appropriate to the individual case.

    †Delete or complete items, as appropriate. Where a failure to comply with the Regulations is indicated further details should be entered, if necessary, overleaf.

Chapter 11
Environmental Installations

What is the purpose of installing a kitchen fan?

Electric ventilating fans should be a must for most kitchens. The preparation of cooked meals is invariably followed by stale smells and unpleasant odours. Steam is also produced resulting in the damaging effects of condensation as it strikes cold surfaces. All these unfavourable conditions may be removed by extractor fans, which, being of small power consumption, are economical to run; the loading may be rated as low as 30 watts.

Fans may be installed in windows but it is preferable to house them in an outside wall, venting directly into the atmosphere and providing a neat flush louvred panel in the kitchen. The unit should be operated during and after the food preparation and, wherever possible, positioned near to the source of smells. Its use will also make for a cool comfortable feeling during spells of humid or hot weather.

What is the major consideration in extractor fan selection?

Correct fan size is a critical factor. This may be shown by a simple example for a kitchen 3 metres (10 ft) by 2.5 metres (8 ft) with a ceiling height of 3 metres.

The total kitchen volume is then equal to $3 \times 2.5 \times 3 = 22.5$ cubic metres. For 15 air changes per hour, the fan must be able to extract 22.5×15 m^3/h, which is 400 cubic metres of air per hour.

Electrical connections to the fan can be obtained from a fused spur unit with wall switch control. Cord switches are preferable for ventilating fans sited near to the kitchen sink.

What are other uses of extractor fans?

Common applications would be for bathrooms and toilets, in particular where these rooms are situated in central parts of blocks of flats, are windowless and have no natural forms of ventilation.

These fans are normally connected to the lighting circuit and start to operate when the light is switched on. The Aidelle Model is fitted with a powerful centrifugal impeller and normally runs for 20 minutes after the light is extinguished. They may be wall-mounted or ceiling-mounted and ducted to a suitable venting point on an outside wall or through the roof. Fitting of an anti-draught flap prevents the entry of draughts when the fan is not in operation.

Some alternative arrangements are

(1)   With natural lighting, the fan timing period can be commenced by operating a door or pull cord switch, including a neon lamp serving as an indicator while the fan is running.

(2)   The fan unit may be fitted in a partition wall between toilet and bathroom so as to vent both rooms.

(3)   Two-speed fan: here control requires a 3-position switch in addition to a separate isolator.

(4)   Inclusion of a standby unit which automatically comes into operation should the primary fan fail.

All equipment and work should comply with the recommendations as given in BS 413:1973 and the relevant requirements for the Building Research Establishment as contained in Digest 170.

In a block of flats, each of the bathrooms is to be ventilated by means of a common extraction system arranged so that, when one of the bathroom lights is on, the common ventilating system will run. What circuit diagram shows the switches and lights in three bathrooms, including a contactor which will control the fan motor?

It should be borne in mind that the supplies to each of the bathrooms and the fan motor must remain separate for metering purposes.

The complete diagram of the circuits is given in Fig. 70. A contactor coil must be connected in parallel with each light point. These coils operate a common contactor lever enabling the extractor fan to be switched on and off from each bathroom.

The fan is connected to the landlord's meter so that its circuit is completely isolated from the bathroom circuits; the latter should all be on the same phase.

(This question, as modified, was taken from a City & Guilds Electrical Installation Examination Question Paper to which due acknowledgement is given. It indicates some of the problems which may have to be encountered when carrying out this type of problem.)

104 *Electrical Installation*

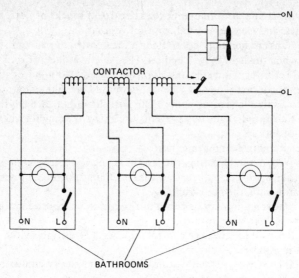

Fig. 70 Ventilation of inside bathrooms

How is it possible to create a comfortable indoor atmospheric environment?

Since surroundings for human comfort depend upon such factors as temperature, humidity and air purity, the obvious answer is *air conditioning.*

This form of indoor atmospheric control — usually by automatic means — has wide application and is an advance on simple ventilation. It is linked with air refrigeration and heating to result in beneficial conditions, which certainly applies to tropical or sub-tropical areas, but such comfortable conditions are increasingly being requested by all societies as demanded by higher standards of living. Contrary to general belief, air conditioning can be arranged for heating purposes during cold periods.

Air conditioning is also essential for various industries when it is required to have a precise amount of water in the atmosphere, or duct washing and clearing of toxic gases or smells.

Large stores and hotels will have a complex central chamber connected to a vast ducting system connected to all rooms. The modern simplified split system air conditioner (Fig. 71) is fitted to individual rooms for commercial and even domestic purposes. These separate units have the advantage of catering for the different needs in various zones.

Fundamentally, all air conditioning units consist of a chamber in which air is brought in to contact with water (which may be in the form of sprays)

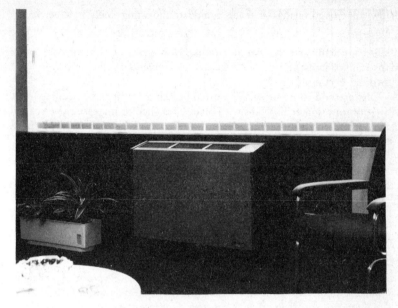

Fig. 71 Myson split unit air conditioner

so as to regulate the temperature and water content of the air. For cooling purposes, an outdoor condensing heat pump draws refrigerating material into a heat exchanger. The base of the unit draws warm air which is discharged inside the room as cooled air.

For heating, air drawn from outside heats refrigerated gas which is then pumped into the heat exchanger to disperse heat into the room.

By operating on the heat pump principle, air conditioning makes for economy and an efficient system of operation.

Following the previous question, what further air conditioning components are available and how can economies be effected?

Essential components to provide comfort conditions by air are dust filters, humidifier eliminators, and silencer arrangements. Due to the drop in air pressure, the largest proportion of fan power is lost in the ducting, partly because of the duct air friction. Such losses can be lessened by reducing the duct diameter wherever possible.

Typical dust filters are electrostatic in action and may be of the bag or roll type. Because of an ability to extract a high quantity of dust, the bag type with its greater fan power is now favoured. In conjunction with coolers the initial cost is high but pays over a long period. Clearly, close attention to good design can make for considerable savings.

How can energy savings be made in industrial heating solely by electric fans?

Fans are usually associated with cooling. However, since hot air rises, ceiling fans can produce the reverse effect. The heat energy at these elevated positions is completely wasted.

One method for re-using this wasted hot air is by fitting, at ceiling level, low-consumption ceiling fans with blades designed to increase the circulation area and thus 'throw back' the hot air to floor or working level. Fan speed should be fitted for adjustments to suit variations in temperature.

With care in design, savings in the electricity energy bill could be as much as 20%, while the fan system is naturally suitable for cooling during the warmer spells of weather.

What is the instantaneous electric shower?

Due to many historical factors, Britain is largely bath-orientated unlike several other countries, but this new appliance may well produce a change in bathing habits. As a relatively recent form of electric water heating, it is perhaps not surprising that there are a number of attractive and modern technical features:

(1) The shower is started by an electronic fingertip control and continues for a full five minutes (or less if required) before being automatically turned off.

(2) A stabilisation control ensures that there is a full flow of water at a fixed temperature. A number of control settings allow the latter to be suitably varied.

(3) An anti-scald device operated by a thermal cut-out is incorporated.

(4) Pressure switches ensure that the elements are switched on only when an adequate supply of water is available.

Thorough earthing facilities are included and showers are to comply to BS 3456. Instantaneous heating is obtained by means of a 7-kilowatt heater, and therefore they cannot be run off the 30 A ring circuit but require a separate 30 A supply.

With the loading of 7 kW on for 5 minutes, the electrical energy consumed for each shower is 0.6 kilowatt-hour, so that for the costing of the running charge, five showers may be had by three electrical units (3 kWh).

Can you supply a solution to the following water heating problem?

(a) 900 litres of water are electrically heated between 11.00 p.m. and 7.00 a.m. The temperature of the water at 11.00 p.m. is 30°C and no water is drawn between 11.00 p.m. and 7 a.m. Find the power of the heater if the water temperature at 7.00 a.m. is 80°C. Assume that 5% of the energy supplied is lost.

(b) If 50 litres are drawn off shortly after 7.00 a.m. and the temperature of the water supply is 15°C, find the resultant temperature of the water and the time required for the heater to raise the temperature of the water to 80°C again. (C & G)

(a) Since one litre of water has a mass of 1 kg,
Mass of water = 900 kg
Temperature rise = 80 − 30 = 50°C
Time = 8 hours
Specific heat of water = 4190 J/kg°C
Therefore
$$\text{Heat gained by water} = \frac{900 \times 50 \times 4190}{10^6}$$
$$= 188.55 \text{ MJ}$$
Since efficiency = 95%
$$\frac{95}{100} = \frac{\text{Input}}{\text{Ouput}}$$
$$\text{Input} = \frac{188.55 \times 100}{95} = 198.5 \text{ MJ}$$
Also
$$\text{Power} = \frac{\text{Energy}}{\text{Time}}$$
$$= \frac{198.5 \times 10^6}{8 \times 60 \times 60 \times 1000} = 6.892 \text{ kW}$$

(The correct changes should be noted for the conversion to kW.)

(b) The mixture of water is made up of 850 litres at 80°C and 50 litres at 15°C, so that resultant temperature is
$$\frac{850 \times 80 + 50 \times 15}{900} = 76.4°C$$
Temperature rise = 80 − 76.4 = 3.6°C
$$\text{Heat gained by water} = \frac{900 \times 3.6 \times 4190}{10^6}$$
$$= 13.58 \text{ MJ}$$
$$\text{Electrical input} = \frac{13.58 \times 100}{95}$$
$$= 14.3 \text{ MJ}$$
Since Power = $\frac{\text{Energy}}{\text{Time}}$

$$\text{Time} = \frac{\text{Energy}}{\text{Power}}$$
$$= \frac{14.3 \times 10^6}{6.89 \times 1000 \times 60}$$
$$= 34.5 \text{ minutes}$$

What is the electrode boiler?

This is an unusual method of heating water and is adopted in situations where large quantities of hot water or steam are required. The Berkely Nuclear Laboratories of the Central Electricity Board of Great Britain (C.E.G.B.) has such a boiler with a loading of 4 megawatts (4×10^6 W).

Generally, three electrodes from a 3-phase supply at low voltage are inserted in an appropriate tank and, since the water acts as a resistor, a heat rise is generated.

Due to the fact that the electrode bare conductors are in direct contact with water, stringent safety requirements are essential. These are set out in the Electricity Supply Regulations and the IEE Regulations for Electrical Installations. Agreement with Water Authorities concerned is also necessary. Because of the setting-up of stray leakage currents affecting telephone lines, the various Water Boards must also be informed. They would normally require to know the position of every point where the system is connected to earth.

The electrical safety regulations stipulate that

(1) Control is required by a linked circuit-breaker with an overload protective device in each conductor.

(2) A device such as an interlock must be fitted to prevent the electrodes from becoming alive when maintenance is being undertaken on the boiler.

(3) An unusual feature is that the metal shell of the boiler must be bonded to the metal sheath of the incoming cable and in turn connected to the neutral conductor. (This arrangement is somewhat similar to the requirements for the p.m.e. system.)

(4) The cross-sectional area of the bonding cable must be equal to the cross-sectional area of the supply conductors.

There is a need for economic use of fuel resources. Is there any way we can still enjoy the benefits which electricity has to offer without depleting the world store of fossil fuels?

Here wave and wind power have their place. However the Middle East and other parts of the globe which have long spells of sunshine are in a special position.

There is a fundamental law of thermodynamics that energy cannot be created or destroyed and can only change in form (we have seen this applied to water heating). Fullest use should be of commercial units which, when positioned on roof-tops, capture solar energy and convert it directly into electrical energy. This energy can then be stored into battery secondary cells for use as required.

Index

Air conditioning, 104–5
Aluminium conductors, 15, 59
Augur, 67

Basic requirements, 1
Boxes, 24, 80–1
Busbar trunking, 47

Cables
    armour, 10
    butyl rubber, 18
    colour, 17, 19
    concealed, 67
    polychloroprene, 34
    precast, 27
    p.v.c., 8
    silicone rubber, 18
Ceiling roses, 85–9
Certificates, 101
Conduit
    bending, 25–7
    classification, 21
    precautions, 24–5

Dimmer switch, 81
Discrimination, 11–12
Diversity, 73
Double insulation, 5
Double-pole fusing, 10

Earth-concentric wiring, 58
Earthing, 13–14, 95–7
Eddy currents, 56
Electricity (Factory) Regulations, 1
Electricity Supply Regulations, 1
Electrode boiler, 84
E.S. lampholder, 84
Extra-low voltage, 12

Fans, 102
Farmyard wiring, 68
Fixings, 37
Flameproof, 29, 69
Flexible cords, 16–19
Fuse, 2–11

Inspection fittings, 23
Instantaneous showers, 106
Insulated conduit
    bending, 38–9
    box entry, 37
    small-bore, 40
Insulation faults, 8–9

Lampholders, 88–9
Light-gauge conduit, 21
Line diagram, 76–7
Loop-in box, 28
Low voltage, 12

M.I. cable bending, 55
Miniature circuit-breaker, 4

Neon indicators, 46–7

Plugs, 72–7, 91
Polypropylene, 34
Prefabrication, 39

Shaver socket, 84
Shock values, 7
Short-circuit, 3
Skirting trunking, 44
Space factors, 21
Spur points, 75
Surge diverter, 56
Switched neutral, 6
Switching, 31–3, 63

Time switch, 91
Trakline, 49
Traywork, 71

Ventilation, 103
Voltage drop, 9

Water heating, 106–8
Wood joist care, 66